elementals

ii. air

volume ii

air

Daegan Miller, editor
Nickole Brown & Craig Santos Perez, poetry editors

Gavin Van Horn & Bruce Jennings, series editors

Humans
& Nature
press

Humans and Nature Press, Libertyville 60030
© 2024 by Center for Humans and Nature

For more information, contact Humans & Nature Press, 17660 West Casey Road, Libertyville, Illinois 60048.
Printed in the United States of America.

Cover and slipcase design: Mere Montgomery of LimeRed, https://limered.io

ISBN-13: 979-8-9862896-3-2 (paper)
ISBN-13: 979-8-9862896-4-9 (paper)
ISBN-13: 979-8-9862896-5-6 (paper)
ISBN-13: 979-8-9862896-6-3 (paper)
ISBN-13: 979-8-9862896-7-0 (paper)
ISBN-13: 979-8-9862896-2-5 (set/paper)

Names: Miller, Daegan, editor | Brown, Nickole, poetry editor | Perez, Craig Santos, poetry editor | Van Horn, Gavin, series editor | Jennings, Bruce, series editor

Title: Elementals: Air, vol. 2 / edited by Daegan Miller

Description: First edition. | Libertyville, IL: Humans and Nature Press, 2024 | Identifiers: LCCN 2024902614 | ISBN 9798986289649 (paper)

Copyright and permission acknowledgments appear on page 120.

Humans and Nature Press
17660 West Casey Road, Libertyville, Illinois 60048

www.humansandnature.org

Printed by Graphic Arts Studio, Inc. on Rolland Opaque paper. This paper contains 30% post-consumer fiber, is manufactured using renewable energy biogas, and is elemental chlorine free. It is Forest Stewardship Council® and Rainforest Alliance certified.

contents

Gavin Van Horn and Bruce Jennings
Gathering: Introducing the Elementals *Series*   1

Daegan Miller
Introduction: Breathe   5

Aimee Nezhukumatathil
Bodies in the Air   8

Sohini Basak
An Elegy with Holes (for the Insects to Come and Go)   9

Andrew S. Yang
On the Poetics and Mathematics of Air   19

Ellen Bass
Saturn's Rings   25

Darran Anderson
How the Sky Was Lost   27

Nicholas Triolo
Castles in the Air   38

Felicia Zamora
On Blessings & Want   49

Sara Beck
We Will Know Ourselves Beloved   51

Michele Wick
Air Space   61

Ross Gay
A Small Needful Fact   70

Roy Scranton
71   *Aura/Error*

Antonia Malchik
85   *Trespassing*

Benjamin Kunkel
92   *Inebriate of Air*

Rita Dove
97   *Ozone*

Báyò Akómoláfé and Daegan Miller
An Invitation to Lose One's Way:
99   *In Conversation with Báyò Akómoláfé*

Craig Santos Perez
109   *American Atmosphere*

Gabrielle Bellot
110   *To Bring Down a Sky*

120   *Permissions*

121   *Acknowledgments*

123   *Contributors*

Gathering: Introducing the Elementals Series

Gavin Van Horn and Bruce Jennings

T hunderous, cymbal-clashing waves. Dervish winds whipping across mountain saddles. Conflagrations of flame licking at a smoke-filled sky. The majesties of desert sands and wheat fields extending beyond the horizon. What riotous confluence of sound, sight, smell, taste, and touch breaches your imagination when you call to mind the elementals? Yet the elementals may enter your thoughts as subtler, quieter presences. The gentle burbling of clear creek water. The rich loamy soil underfoot on a trail not often followed. A pine-scented breeze wafting through a forest. The inviting warmth of a fire in the hearth.

This last image of the hearth fire is apropos for the five volumes that constitute *Elementals*. The fire, with its gift of collective warmth, is a place to gather and cook together, and not least of all a place that invites storytelling. And in stories the elementals can be imagined as a better way of living still to be attained.

The essays and poems in these volumes offer a wide variety of elemental experiences and encounters, taking kaleidoscopic turns into the many facets of earth, air, water, and fire. But this series ventures beyond good storytelling. Each of the contributions in the pages you now hold in your hands also seeks to respond to a question: What can the vital forces of earth, air, water, and fire teach us about being human in a more-than-human world? Perhaps this sort of question is also part of experiencing a good fire, the kind in which we can stare into the sparks and contemplate our lives, releasing our imaginations to possibilities, yet to be fulfilled but still within reach. The elementals live. Thinking and acting through them—in accommodation with them—is not outmoded in our time. On the contrary, the rebirth of elemental living is one of our most vital needs.

For millennia, conceptual schemes have been devised to identify and understand aspects of reality that are most essential. Of enduring fascination are the four material elements: earth, air, water, and fire. For much longer than humans have existed—indeed, for billions of years—the planet has been shaped by these powerful forces of change and regeneration. Intimately part of the geophysical fluid dynamics of the Earth, all living systems and living beings owe their existence and well-being to these elemental movements of matter and flows of energy. In an era of anthropogenic influence and climate destabilization, however, we are currently bearing witness to the dramatic and destructive potential of these forces as it manifests in soil loss, rising sea levels, devastating floods, and unprecedented fires. The planet absorbs disruptions brought about by the activity of living systems, but only within certain limits and tolerances. Human beings collectively have reached and are beginning to exceed those limits. We might consider these events, increasing in frequency and intensity, as a form of pushback from the elementals, an indication that the scale and scope of human extractive behaviors far exceeds the thresholds within which we can expect to flourish.

The devastating unleashing of elemental forces serves as an invocation to attend more deeply to our shared kinship with other creatures and to what is life-giving and life-nurturing over long-term time horizons. In short, caring about the elementals may also mean caring for them, taking a more care-full approach to them in our everyday lives. And it may mean attending more closely to the indirect effects of technological power employed at the behest of rapacious desires. Unlike more abstract notions of nature or numerical data about species loss, air measurements in parts per million, and other indicators of fraying planetary relations, the elementals can ground our moral relations in something tangible and close at hand—near as our next breath, our next meal, our next drink, our next dark night dawning to day.

For each element, the contributors to this series—drawing from their diverse geographical, cultural, and stylistic perspectives—explore and illuminate practices and cosmovisions that foster reciprocity between people and place, human and nonhuman kin, and the living energies that make all life possible. The essays and poems in this series frequently approach the elements from unexpected angles—for example, asking us to consider the elemental qualities of bog songs, the personhood of rivers, yogic breath, plastic fibers, coal seams, darkness and bird migration, bioluminescence, green burial and mud, the commodification of oxygen, death and thermodynamics, and the healing sociality of a garden, to name only a few of the creatively surprising ways elementals can manifest.

Such diverse topics are united by compelling stories and ethical reflections about how people are working with, adapting to, and cocreating relational depth and ecological diversity by respectfully attending to the forces of earth, air, water, and fire. As was the case in the first anthology published by the Center for Humans and Nature, *Kinship: Belonging in a World of Relations*, the fifth and final volume of *Elementals* looks to how we can live in right relation, how we can *practice* an elemental life. There you'll find the elements converging in provocative ways, and sometimes challenging traditional ideas about what the elements are or can be in our lives. In each of the volumes of *Elementals*, however, our contributors are not simply describing the elementals; they are also always engaging the question, How are we to live?

In a sense, as a collective chorus of voices, *Elementals* is a gathering; we've been called around the fire to tell stories about what it means to be human in a more-than-human world. As we stare into this firelight, recalling and hearing the echoing voices of our living planet, we stretch our natural and moral imaginations. Having done so, we have an opportunity to think and experiment afresh with how to live with the elementals as good relatives. The elementals set the thresholds; they give feedback. Wisdom—if defined as thoughtful, careful practice—entails conforming to what the elements are "saying" and then learning (over a lifetime) how to better listen and respond. Pull up a chair, or sit on the ground near the crackling glow; we'll gaze into the fire together and listen—to the stories that shed light and comfort, to the stories that discombobulate and help us see old things in a new way, to the stories that bring us back to what matters for carrying on together.

Introduction: Breathe

Daegan Miller

In the beginning was the word... but this can't quite be right, for a word is the child of sound, waves rippling the air, and the desire to speak, a desire we call "inspiration" and whose Latin root, *spīrāre*—also the root of the word *spirit*—means to breathe.[1] In the beginning, then, long before the first word cried its infant sound into the world, before the deep breath of inspiration, there was air. The pre-Socratic philosopher Anaximenes held that air was the foundation of everything that was, is, and will come to be, that, "just as our soul, being air, holds us together and controls us, so do breath and air surround the whole cosmos," that air, as Cicero understood his predecessor's thinking, was godlike, "infinite and always in motion."[2]

Infinite and always in motion, constituent of everything—we need not call ourselves disciples of Anaximenes to feel the weight of his truth: air is unutterably old, and its spirit carries with it the story of everything.

But what that story is, it is impossible to know. For try as we might to hear the voice of the wind, or measure with precision the increasing amount of one chemical compound compared to another, or augur whether the breeze blows fair or foul, our aeromancy can only ever fall short, for air is a trickster element. Everywhere, yet impossible to define: Is air the same as oxygen? If so, what was there before cyanobacteria, the first photosynthesizing form of life, destroyed the planet's methane-rich atmosphere about 2.5 billion years ago by exhaling enormous amounts of oxygen, triggering the mass extinction of anerobic life, and ushering in the world that has made our existence possible? Is air no longer air once it carries elements of water or earth or fire's ashen offerings? When precisely does air become something else, and after how much aeration does something else become air? When and where does air begin and stop, and can you point to it, or, perhaps more to the point, point to somewhere, sometime that it isn't?

This much we know to be true: There is no life as we live it without air. There is no cry of birth without air, and we call those ocean zones dead when vast effluent-loving algal blooms suck all the oxygen from the water. Polluted air is one of the leading causes of premature death worldwide, the planet's preternaturally hot air is proving deadly, and strangulation has become one of the favored weapons of American police. "I can't breathe" said Eric Garner to his killer, Officer Daniel Pantaleo of the New York Police Department—Garner, a man who, Ross Gay reminds us in "A Small Needful Fact" elsewhere in this volume, planted some of the many plants that make the air sweeter for us all.

There is no creativity without air, no creation.

For all its elemental power, air is also strikingly vulnerable. It is easily poisoned—as I write this in the summer of 2023, Canada is on fire, and everything that respires, from North America to Europe, chokes on the aerosolized boreal forest carried on the wind—and it heats readily: nearly every one of my forty-three years has been the hottest year on record, and given the complexities and various feedback loops, known and unknown, of climate change, there's little hope of cooling things down any time soon.

What, then, to make of a thing as ineffable as air? What, if anything, can it teach us?

Perhaps those are the wrong questions. Better, I think, to ask how can air be let into the things we make.

The web is the preferred metaphor for ecology, and its purpose is to illustrate the truism that everything is connected. But the metaphor errs, for if it's true that the strands of a web occasionally form nodes of interconnection, it is also true that the vast majority of a web is made up not of silk, but of air. All those spaces in between the connecting threads: they are not empty but full of air and all that it carries. Air spaces. Spirit spaces. Breathing spaces.

Each of the writers in this volume weaves a web of words with great precision, and although their webs take varying shapes—a gonzo journey to the heart of the Anthropocene; a tale of brokenness, birth, and breath; a history of flight; and an avant-garde play bending time and space and character—the geometry of each is tightly constructed to ventilate word and thought. When they can breathe, letter strings grow vivid and the page becomes a habitation for the reader, for air spaces are places where the imagination thrives.

Only the air knows its complete history, and only it can know to where it ultimately blows. We may live in anxious times, but art and good, critical thought are evidence that inspiration, that spirit, yet lives. And so, with the poet Ovid, the writers in this volume release their words:

My soul would sing of metamorphoses.
But since, o gods, you were the source of these
bodies becoming other bodies, breathe
your breath into my book of changes: may the song I sing be
seamless as its way
weaves from the world's beginning to our day.[3]

notes

1. *Oxford English Dictionary*, s.v. "inspiration (*n.*)" and "spirit (*n.*)," accessed online via the University of Massachusetts Amherst library.
2. None of Anaximenes's original texts survive, and what we know of him, we know through his influence on others. See S. Marc Cohen, Patricia Curd, and C. D. C. Reeve, eds., *Readings in Ancient Greek Philosophy: From Thales to Aristotle*, 2nd ed. (Indianapolis: Hackett Publishing, 2000), 12, 13.
3. Ovid, *The Metamorphoses*, trans. Allen Mandelbaum (New York: Alfred A. Knopf, 1993), 7.

Bodies in the Air

Aimee Nezhukumatathil

If you want to talk, just say the word, hummingbird.
I want to know when you will flic-flack from crepe myrtle

or beautyberry bush. In Morocco's Erg Chebbi desert,
spiders cartwheel down sand dunes and how can you

not smile when you learn there's an Italian insect
called *graminicola* that somersaults in the forest?

Splash tetras leap from the river onto leaves
fin-stuck, just long enough for eggs to grow

then swell into a very fine fry. Salamanders backflip
over the face of a rock to avoid birds, whip their tails

up and over, throw their long backs till tiny forelegs reach
for the clouds. So when you finally speak it into the air

and let it leap off—remember it's the only thing
I've ever wanted from you. I want to see your words

hover in the air—watch them be the only feathered thing
on this blue planet that can zip backwards and even upside down.

An Elegy with Holes
(for the Insects to Come and Go)

Sohini Basak

The smell in the   says you cannot go back. The color says it's a memory. The sound is the gray-green of a June night from 1994 and a blinking dot red on an old tape recorder that's wound out the lullabies it memorized the last hour. Nighttime is a single word with double *t*s to mark time. She is four and safe next to her mother. It is summer, it is June, framed first by the small house, then the square window, then the mosquito net tied taut over their bed. Inside, the last words of an old story are interrupted by the arrival of a featherlight firefly. Invisible wings, blinking green. And her mother stops: first the story, then the motion of the ceiling fan. So that the firefly doesn't get hurt, she whispers as if her worry too could hurt the firefly.

Hurt by   ?

No, the blades of the fan that move the   . Can we adjust to the heat because we're sharing our room with the firefly tonight?

Then the focus is on the wind on the window, on how even on a hot humid night the outside makes the inside so cool and even beautiful and dark and full of small movements, slivers of the moon, and scents and sounds faint yet constant like waves in the   as the child, the mother, and the firefly breathe in and breathe out the same atmosphere.

Then a child's worry: who else could be inhabiting this   ? Ghosts and gods and angels and the invisible spirits moving there.

What's behind the curtains?   The world.
And in front of it?   You, me, our sleep.
And there—what's that all over?   All over where?

She means to say all over and all around the gauze of the mos-
quito net and the lace of the curtains—the haunting in the fibers
of the world and under it and on it and over it and between the
space, between all things breathing and still. She doesn't know
the words, the verbs or the prepositions yet. She does not know
how to grasp the words to pinpoint her large feelings of fear and
revery, but she tries. She doesn't know it yet, but she will spend
her life trying to pinpoint moments such as these with words. The
way the   felt on the night, cool over her sweaty temple. She
will never be able to describe or retell that thing in the   . A
switch flicked on and off, then on and off again. Something moves
through the   .
Why, that's just   .

Another summer, or all summers until the weather turned
for good, pink-yellow evenings lined by the flight of migra-
tory birds. February, month of departures, month of collective
wings leaving the jute-factory town in the shape of a V, slicing
through the   .

Another summer, or all summers until the wind patterns shift-
ed for good, she waited for the afternoon storms. She waits for her
mother to return home before the afternoon Kalboishakhi began.
One day, before her mother returned, the   thickens with
black clouds and several of the flowerpots on the rooftop dislodge
in the wind. Minor terra-cotta trees roll around. One breaks, an-
other drops off the ledge and shatters on the street. Then thunder,
then lightning, then lashing cold rain. She thinks her heart would
evaporate then.

It's good that it rained, her mother says, returning two hours
later. It means we're on track.

She wants to tell her how it was terrible, that she felt betrayed and abandoned, and how could she be so unfazed, but she asks, For what, what are we on track for?

For rice, for fruits, for vegetables, for the fish and all our seasons.

She takes care that day to chew her food slowly. She takes care that day not to swallow too much   .

And now this.

The smell in the   says it's roadworks day. Years later, and it is still February. It is her second week in a new city far away from the hometown, it is farther away from the other city she's left behind, it is tar-colored 11 a.m. and the sun is already overhead saying, Is it the day to cut down trees and pour concrete over the earth and build more nests for you lot? The color in the   says it's smoke it's development it's progress it's for profit only it's election season. At least with the roads repaved, the problem of waterlogging has been discussed, dismissed. The color of the   says it's spring. It is March. It is almost April. The nor'westers should be here any minute. The icon on the app says there is a 95 percent chance of precipitation at 16:45. It would be foolish to do laundry today—

She goes cloud watching every afternoon. Late July, early 2000s, the sky an impossible Pantone 2386 C blue. Congregating like sheep, her vacant thoughts cloud over her anxieties and worries. Some days ought to be vaporous, vapid. After school, alone in the house, she dials up thick curtains of daydreams. Now she is green, now she is fire, now she can fly. It is no secret that she loves being an   head.

First, fear.

   Then, friendship.

      Finally, betrayal.

         Somewhere in between (under and over and always)—

In 2015, the   around the capital changes for good. She lives alone now in a room on a rooftop, growing mint and marigold to calm her nerves and calm down the bees. In her office cubicle, she dreams about tending to a moon garden. Pale, white flowers, silver leafy beds. But if they are rooted to heavy pots, how to bring them indoors when the dust storms come knocking, swirling sand with   ? What she once feared, now she learns to witness, with attention, with caution. She dreams of a year when all evenings are breezy. When every day is sunny and every night sweet and balmy. What she gets or what she makes or what she invariably votes for is various degrees of gray. She wakes one October morning to the smell of burning leaves. No pile of fire on her street or the neighborhood, but it's denser this smell. It sticks to her clothes and her towel drying on the rooftop, and it even consumes the mint and the marigold. Over the next weeks, months, years, the marigold wilts. The   turns thick, yellow, sick. She doesn't know it yet, but this cough will haunt her, slink back in with the poisoned draft through the open window.

And now? Well, now it's still March. It's past 17:00, the middle of mosquito hour. The windows are shut and dry and the leaves outside are few and still. The quiet before a summer storm. The skyline is dust-colored buildings lined by darkening clouds. Not black and ethereal like from ten years ago, but harsher somehow. Or nostalgia is blooming unscientific thoughts again. The color of the   is colored by barbets trying to sing over the whirring machines building an apartment after clearing off a chikoo garden. On the streets, small yellow-green seeds fall over the concrete and as the hours deepen the birds go deaf—

Cut back to 2008. It's only a case of postnasal drip, nothing that a little spray bottle and staying away from construction sites won't cure. Cut to 2010. Only a case of tachycardia that acts up now and then. Jump to 2016 but stock up on those N-95s first. Meanwhile, larvae swell up and flourish in the moist   . Rest and recuperate after a bout of life-altering dengue. Get up and jump to 2020 and mask up to things that are   -borne. Slowly now walk (through burning hoops) to 2022 and acquire a full set of food and environmental allergies. Cut to 2023, no cut 2.3 million trees in 2023. We were   -born and still counting. All, because of the   .
The ribs the lungs the   -ways the alveoli the mucus the nose the mouth the pores the cells dividing or dying. Anxiety spirals like a cyclone forming over the Bay of Bengal. A wall of worry upwells from the depths. No longer space but smog between us. No, really, what's that we can but shouldn't see?

What's that in the   ?

Particulate matter and ozone and wingspan and nitrogen dioxide and drone strikes and migrating monarchs and asbestos and lead and helicopter blade and aurora borealis and hurricanes and carbon monoxide and tumbleweed and SUVs and space debris and pollinator anosmia and cirrocumulus and echolocation at 100,000 Hz and comet tails and methane and volcanic ash and fiberglass and hailstones and hot-air balloons and murmuration and pathogenic droplets and I-giga-heart-you radio waves and turbines and the spectacle of purple-orange sunsets and ISS and ventifacts and PAHs and infrared and dandelion clocks and private jets and tidal waves and white phosphorous and everything is coal or cobalt colored and the sound is that of our greed drilling right through the hearth of our one home—

Mighty dust motes, don't leave us bereft—

And now? Now, the kite is tired after mourning for thirteen days for the chikoo garden that has been razed to red dust. Perched over a dish antenna, she screeches at dusk and flies off to another part of the city.

The color of the   says it is another memory. She is eleven and talking about the Coriolis effect to a group of strangers at a school fair. She is in love with ocean currents and wonders whether she could become a meteorologist. Next to her, her classmate is talking about air pollution and saying, We're a bunch of strange creatures thinking we invented maths, but we've got it all wrong! We're adding to what should be empty (the   , the sea) and we're deducting (forest cover, riverbeds, coastlines) from what should be left alone whole! She has forgotten the classmate's name and face and voice, but she remembers him sometimes for what he had said at that school fair.

And now… the sun icon is replaced by a clouded moon icon. Beneath the mass of pixel vapors, three tear- or raindrops glow suspended. Sixty-five percent chance of a summer shower at 20:30. Time to prepare by charging the phone, keeping a battery torch close at hand. These things, these patterns of the   , can go out of hand these days and so, as the wind picks up, there is that urgent cold smell in the   . So, one by one, the small potted plants are brought indoors from the narrow and precarious balcony. No risking broken china and shattered terra-cotta this early into her move. Sorry you will miss the season's first rain. Her apologies doled out to the leaves of the new mint and the new marigold.

As childhood vanished into the thin   , so did regular Kalboishakhi, sparrows, swarms of honeybees, candles during power cuts, drongos and dragonflies in August, silly dreams of being a caped crusader, a weather girl, the smell of winter rain, migratory birds, deep red kolaboti blooms, and within their leaves, on pitch-dark nights, a family of fireflies.

The color of the   screams orange gold black blazing. The smell in the   says there are 340-plus forest fires. What's that in the   ? What's burning? All that's solid melts into—
  Empires fall and altars are extinguished but some fires burn on.
        Sanctuary lamps, mud volcanoes, goddesses of fire,
                memorials for peace.
        In places like Jharia, where the mine fire has not stopped
                burning in one hundred years.
  And now this, our earth, the   above us, an eternal flame.
        All that's burning deposits onto the lungs.

Windows rattle, curtains shiver. Last year, and the year before last, sudden summertime cyclones and incessant rain created water-logged streets, a rareness in the   . Loss and displacement and disease and submergence, the promise of progress, of rehabil-itation unhinged with new patterns of heat and cold. A stirring. A stagnation. A hollowness gathers in the sky or the chest. A heaviness. Part factory, part moisture. An unwanted embrace.

Something moves through the ⠀⠀⠀⠀. Birds fall from the sky like portents. Large bees, lying dead on the staircase. Welcome gale, welcome disruption, welcome new strains of virus, welcome we're-running-out-of-names-for-cyclones. News cycle quickly changing, spawning like the quickening life cycle of mosquitoes. Without the ⠀⠀⠀⠀, impossible to see, impossible to hear, to smell, to be. But try, please listen. Listen to the great body of the earth. Listen to your own breath trapped like a moth in a butterfly net. The breath of your loved ones. The gurgles and the rumbles and the hacks and the wheezes that should not have been there. An arrhythmia of the planet as of your own heart. An acknowledgment of a collective ailment, steadily advancing.

The failure to express, to build with words, that one sentence about that which spreads over from the outside to the inside, that which we all share, the porosity of vessels, cell density, humidity or a window that separates the inside from the outside a window that connects your home to the barbet's the wasp's the hummingbird's the snowy owl's the moss spider's the kakapo's the tree frog's the tiger butterfly's the great Indian bustard's the carpenter bee's the cicadas' hitting hard the glass. And yet harboring hope like a craving for seasonal rain that will not go away, like moisture miraculous beneath the topsoil of our prayers for the unseasonal heat to dissipate. An open window so forgiving it regulates our desires, sifts wants from needs.

We feel cold so we layer up we insulate our hearts we feel thunderstruck smokestruck sunstruck moonstruck and so we layer up in PVC and CFCs we hide behind HEPA filters and we line up little wet bulbs on our trendy urban heat islands and instead of curtains stitch together giant gazebo tents out of polypropylene.

If this mouth is a window, how do we let the winged-one out—

Some wind, some dust swirling in the   . Sudden gusts, the minor gods are playing cards again. A movement for a haunting. It's 21:45 and something made of glass shatters next door. A tiny neem flower flies in. The rumbling of clouds, or heavy machinery on the highway. Time to unplug the microwave, the computer, the lamp. The screen lights up the white petals for a second and an elegy unspools. A star wilts in the face of a make-believe dark night. The lights go out. The lights go out again. Time to spray mosquito repellent and pray no one else inhales it in large quantities. Time to open the windows and to let a cool breeze come in. Come, come inside, be our guest, empires of dust.

And what of the wounded trees—
Inconsolable the breeze—

A crack. A streak of lightning. Then the focus is on the wind on the window and how talking about the weather only makes one small in comparison to the melting mountain range. Confounding speech with action hiding somewhere beneath the tip of an iceberg, shifting like English-language definitions and international goalposts. An array of graphs, commas, question marks, and hot   , but still no rain. An unclean smell in the   like stale longing or worry smoking up like signals. Save our souls, help

our ghosts, but what if the vessel is contaminated? An old smell in
the   as if it rained in the next town. As if it will never rain
here again. The stench and sound of the generator now making
the   heavier. Two bats flying by, overhead. Then darkness,
then fire, then time hovering, like flies. Then wind, dry leaves fall-
ing on concrete, the sound of empty rain.

On the Poetics and Mathematics of Air

Andrew S. Yang

A ir is equivocal. It quavers between the significantly ethe-
real and elusively substantial. Air is really there, but not
really there—it is physical but more or less intangible.
When air is recognized as something, it is almost always as a
medium in which other things happen like clouds, pollen, conta-
gion, or music. Consider Bashō—

Spring air—
woven moon
and plum scent[1]

The air is the aroma that it carries forth and the moon that it
suspends like a bright round button. The air is an invisible vehicle
and substrate.

This past summer, the air stung. Doing yardwork outdoors
made me hoarse; I coughed a little, then coughed more. The noon
sky was dusky and the sun oddly orange, akin to a persimmon on
a gray blanket. Micron-sized soot was traveling by air across thou-
sands of miles, coming from the wildfires that seared Canadian
forests in conflagration. Consider Issa—

forming the year's
first sky...
tea smoke[2]

He reminds us that fire does more than inhabit air—and in fact builds it. The air's oxygen feeds the flames, and those flames transform black spruce and fir into hundreds of millions of tons of carbon that remake an ever-hotter atmosphere.

It is graduation day at Harvard University, and students in caps and gowns walk with aplomb. A person stops a few of them and hands each a seed:

> Interviewer: Hold on to that for a second. Imagine that I planted that in the ground and a tree grew. And here is a piece of that tree [the graduate is then handed a heavy log]. Now, where did all that stuff come from?

The graduates offer a variety of answers—water, soil, minerals, and nutrients from the ground.

> Interviewer: Now, what would you say to someone who said to you that most of the weight of a tree came from the carbon dioxide in the air?

> Graduate A: I would say I have no idea, I would have to think about that.

> Graduate B: I would say that is very disturbing, and wonder how that could happen.

> Graduate A: That would be hard to believe because carbon dioxide is, well, it's a gas, and it doesn't seem intuitive that you can take on mass by taking in a gas.

"It is a very strange idea," reflects the narrator, "that somehow the air which they view as nothing, as weightless, as insubstantial somehow makes a tree, a giant tree, that weighs several tons."[3]

Every breath these graduates exhaled was carbon dioxide, from which the Canadian spruce and fir grew their big, flammable bodies. This is a puzzle for aesthetics—how to conjure sensation and comprehensibility from an invisible that is everywhere. The air is heavy with our aspirations, but more and more it feels like we are earthlings without an Earth. How can the intangible everything become more than nothing?

> The atmosphere is not a perfume, it has no taste of the distillation, it is odorless, / It is for my mouth forever, I am in love with it, / ...I am mad for it to be in contact with me.
> The smoke of my own breath, / ...My respiration and inspiration, the beating of my heart, the passing of blood and air through my lungs
> —Walt Whitman, *Song of Myself*[4]

There is no air without Earth. The Cambridge and the Oxford dictionaries converge on almost identical definitions of air: the mixture of gases that surrounds the earth and that we breathe.[5] But in capturing air as both a geophysical phenomenon (that we recognize as sky or space) and an embodied engagement (that we experience as breath), a duality emerges. Air is at once distantly external and intimately internal—an extended vault of atmosphere above us and just as equally a quiet sigh within us. It isn't obvious how to make sense of the Janus-faced disjunct of an air that is here and there but somehow nowhere in between.

Add to this the fact that at this moment you are likely inhaling a molecule of Walt Whitman's last breath, taken on March 26, 1892, in Camden, New Jersey. It is as preposterous and unnerving as it is true.[6] Poetry can have that quality, as can physics; there is a seductive wonder in sharing the smoke of Whitman's own breath across a century, just as there is a sublimity to one among his last 25 sextillion molecules of respiration circulating in the troposphere for generations, only to enter your body right now.[7]

It feels impossible to make full sense of Whitman's physical breath becoming our own, just as we are unable to properly reckon the continuum of the air teeming between the edge of outer space and our nostrils (much less the innumerable bubbles roiling in the oceans or trapped miles deep in Antarctic ice). One might think of air as a hyperobject—something that the literary scholar Timothy Morton has proposed to describe entities that are "massively distributed in time and space as to transcend spatiotemporal specificity" and evade direct sensation.[8] However, if we take the earthliness of air seriously—how it is held tight to the planet by its gravity, disappearing at an altitude of some one hundred kilometers—it turns out to be quite local, the thin skin of a blue marble.[9]

Air may be less a hyperobject and more a basis for what Immanuel Kant called the mathematical sublime. The sublime emerges when nature's raw power and scale exceed comprehension by our senses or our imagination. The mathematical sublime, in contrast, is generated through our intellectual abilities, which, through math and measurement, can tame overwhelming magnitudes by representing them numerically.[10] In this way, our "supersensible faculty" of reason can subdue the fathomless and unify phenomena that appear to be in disparate contradiction otherwise. The false dichotomy of sky and breath can dissolve into a seamless atmosphere, while 25 sextillion molecules of air transfigure from an insensible plentitude into a conceivable magnitude. We can even conclude that there are more molecules of air in a single breath than breaths of air held within Earth's atmosphere. What is both exceedingly small and extremely numerous becomes comprehensible, and maybe even more bewildering in that sublimity.[11]

Consider how for every million molecules of air in the atmosphere right now, only 416 of them are carbon dioxide—in scientific parlance, 416 parts per million (ppm). This minute and closely monitored quantity is a measure of our climate past and future, its continuous increase over the past 250 years a consequence of our unremitting combustion of wood, coal, and oil. If we can conjure calculations to represent any magnitude that nature has to offer, then the pleasure of the mathematical sublime arises from subduing the insensible vastness of scale through reason, creating domesticated versions of unruly realities. As a consequence, parts per million has become the underwhelming currency of potential climate catastrophe, a miniscule measurement that highlights the enormous potency of carbon dioxide in one sense while diminishing it in another.

Whether it is viral pandemics or dark matter or global warming, we wrestle with the paradox that phenomenal immensities are generated by the diffuse, diminutive, and often imperceptible. Perhaps 416 parts per million represents a kind of enigmatic sublime, as persistently unassuming as it is profound. Our aesthetics struggle, understandings wobble.

We do so much with an air of unknowing. The atmosphere is not a perfume, but neither is it arithmetic. What is left but to try to catch our breath?

notes

1. Lucien Stryk, *On Love and Barley: Haiku of Basho* (New York: Viking Penguin, 1985), 31.
2. David G. Lanoue, "Haiku of Kobayashi Issa," http://haikuguy.com/issa/search.php.
3. Annenberg Learner, "Lessons from Thin Air," 1997, https://www.learner.org/series/minds-of-our-own/2-lessons-from-thin-air/.
4. Walt Whitman, "Song of Myself" (1892 version), Poetry Foundation, https://www.poetryfoundation.org/poems/45477/song-of-myself-1892-version.
5. Cambridge Dictionary Online, s.v. "air," https://dictionary.cambridge.org/dictionary/english/air.
6. The calculation and conclusion are based on a classic physics example often called Caesar's last breath. It is an example of a Fermi problem, named after the physicist Enrico Fermi, who frequently used mathematics to quickly estimate complex questions. The physical principles and calculations just as easily apply to Walt Whitman or any number of people, and not just last breaths. An explanation of the calculation can be found here: "Only a Breath Away aka 'Caesar's Last Breath,'" *A View from the Back of the Envelope* (blog), http://www.vendian.org/envelope/dir2/breath.html.
7. James Lloyd, "Are We Really Breathing Caesar's Last Breath?" *BBC Science Focus*, July 12, 2017, https://www.sciencefocus.com/planet-earth/

are-we-really-breathing-caesars-last-breath.

8. Timothy Morton, *The Ecological Thought* (Cambridge, MA: MIT Press, 2010), 130.

9. "The blue marble" refers to a photograph of Earth that was taken on December 7, 1972, twenty-one thousand miles from the surface, by the crew of Apollo 17 on their mission to the moon. In it, Earth is photographed in its totality, "floating" in the darkness of space and offering a profound sense of locality to the Earth and everything upon it.

10. Kant describes the mathematical sublime starting off as "a feeling of displeasure, arising from the inadequacy of imagination in the aesthetic estimation of magnitude," which then turns into a sense of sublime satisfaction that "awakens the feeling of a supersensible faculty in us." See James Creed Meredith, *Kant's Critique of Aesthetic Judgment* (Oxford: Clarendon Press, 1911), 33; and Immanuel Kant, *Critique of the Power of Judgment,* trans. P. Guyer and E. Matthews (Cambridge: Cambridge University Press, 2000), 134.

11. Kant argued that "the sublime is that in comparison with which everything else is small" and thus "absolutely great." Immanuel Kant, *Critique of the Power of Judgment,* trans. J. H. Bernard (New York: Hafner Press, 1951), 87. The absolutely great is most often thought of in terms of three-dimensional size; however, we should consider size as a matter of abundance as well, since the key for the mathematical sublime would appear to be the estimation of any magnitude that is otherwise insensible to our aesthetic capacities, be that large, small, or numerous.

Saturn's Rings

Ellen Bass

Last night I saw the rings of Saturn
for the first time, that brilliant band
of icy crystals and dust. Mirrors
shepherding the light, collecting it
like pollen or manna
or pails of sweet clear water drawn
from the depths of an ancient well.
The gleam poured through my pupils
into this small, temporary body,
my wrinkled brain in its eggshell skull,
my tunneling blood, breasts that remember
the sting and flush of milk.
Saturn, its frozen rings fire-white,
reflecting the sun from a billion miles.
Maybe there's a word in another language
for when distance dissolves into time.
How are we changed when we stand out
under the fat stars of summer,
our pores opening in the night?
The earth from Saturn is a pale blue orb,
smaller than the heart of whomever you love.

You don't forget the poles of the earth
turning to slush,
you don't forget the turtles
burning in the gulf.
Burger King at the end of the block
is frying perfectly round patties,
the cows off I-5 stand ankle-deep
in excrement. The television
spreads its blue wings over the window
of the house across the street
where someone's husband pressed a gun
against the ridged roof of his mouth.
This choreography of ruin, the world breaking
like glass under a microscope,
the way it doesn't crack all at once
but spreads out from the damaged cavities.
Still, for a moment, it all recedes.
The backyard potatoes swell quietly,
buried beneath their canopy of leaves.
The wind rubs its hands through the trees.

How the Sky Was Lost

Darran Anderson

O n a summer's evening in 1969, a family on an island at the edge of Europe is gathered around the television. More than six hundred million people, a fifth of the world, are watching the first moon landing. The grandfather of this family, *my* family, a dapper old gentleman called Jimmy Doherty, is boycotting the event. Instead, he stares out of the window at the darkening fields, saying to himself, "No good will come of this."

For every intrepid explorer, there is a reticent sage warning against the intrusions of excursions. Once a technology becomes ubiquitous, we often forget the fear and hostility it initially aroused.

In Victorian times, for instance, there was a press-fueled panic about train travel. Humans, it was claimed, were not built to travel so fast. Even the *Lancet* succumbed to this "railway mania," claiming that traveling by train resulted in "cerebral disturbance," "affection of the nervous system," "spinal softening," "apoplexy," "violent inflammation" of the eyes, and so on.[1] It is easy to mock the naïveté involved, but it is also easy to overlook the wisdom of trepidation.

From an early age, I was enchanted by the sky. I grew up a mile from where Amelia Earhart landed after her transatlantic crossing, a field sanctified by her alighting. *Earhart* was a magical word to me then. I read everything I could about aviation and its adventurers. And yet I was also terrified of it. One reason for this was the menacing, omnipresent drone of military helicopters over our neighborhoods in Troubles-era Northern Ireland—it felt like everything was under surveillance by almost supernatural entities. Another reason came as the result of listening in to a family friend recounting a plane crash he'd survived in which his wife and many others had been killed. A domestic flight had, in seconds, turned into a bloodbath. Fear and fascination come conjoined, societally and personally. The anxiety and fascination I have felt ever since hearing that story, during hundreds of flights, was there from the beginning of human-powered flight itself, and when that vital, symbiotic connection between them is broken, something crucial to humanity is lost.

To occupy a place in the sky, to dream of flight, is so intoxicating that it has naturally stirred up puritan disdain. At first, this was a case of hubris meeting nemesis, from Icarus flying too high to the legendary English king Bladud coming to an abrupt end when his spells and feathers failed him. With the advent of Christianity, it became an issue of sacred trespass, as the Tower of Babel had been. No one could intrude on the realm of God and his seraphim, cherubim, and archangels. When Simon Magus and St. Peter faced off in a magician's battle in front of Nero, the former made the error of boasting that "he would fly up to heaven since the earth was not worthy to hold him," only to have St. Peter call on the invisible angels who were raising him to let the heretic fall crashing down onto the Via Sacra.[2]

What was really being punished was curiosity, just as curiosity about the forbidden fruit of knowledge led to Adam and Eve's expulsion from Eden. There were notable exceptions, reserved for elite patriarchs called up into heaven or a handful of saints possessing the miraculous ability to levitate (e.g., St. Francis, St. Mary of Jesus Crucified), one of whom, Joseph of Cupertino, was retrospectively designated the patron saint of aviators. Followers would gaze up at basilica ceiling paintings, with heaven re-created in ingenious trompe-l'oeil (Andrea Pozzo's Church of St. Ignacio, for instance), assured that if they lived an obedient pious life, they might be admitted to the afterlife. If, however, they intruded upon the heavens this side of death, they would be, like Milton's proud and defiant Satan, "hurled headlong flaming from the ethereal sky."[3]

It is worth considering whether there might be other reasons for this taboo, aside from a jealous God and the hierarchies of his ambassadors on earth. By its nature, flight would be, and is, dangerous and transient. For all the tales of floating cities in the likes of *Gulliver's Travels*, it is essentially a liminal space of transition rather than one that could be inhabited in any substantial sense. It also creates a perspective problem. Traveling by land or sea, humans can gradually experience and acclimatize to the surroundings. With flight, especially above cloud cover, they enter an uninhabited realm, without topographical or cultural cues, and then potentially descend into vastly different environments and cultures than their origin point. However pleasant the artificial environment might be en route, or how seemingly miraculous the technology that allows it, there may well be a cost to this distortion of our relationship to space and time.

Over time, myth became stricture, and cautionary tales became condemnations of transgressive sin. One of the key accusations leveled against witches, for example, is that they could fly. As these denunciations were patently absurd, especially when presented alongside the often-powerless accused, prosecutors claimed they occurred by means of shape-shifting and under cover of night. In medieval Ireland, Dame Alice Kyteler was accused of witchcraft but escaped abroad, so they condemned her maid Petronilla de Meath instead. Among the many accusations was one that Kyteler and de Meath, with "a secret ointment," "impregnated" a length of wood in order to fly upon it, "carried to any part of the world without hurt or hindrance," alongside claims of murder and demonic copulation.[4] After being whipped, de Meath was burned alive. However disingenuous the indictment, mastery of the skies ranked among the most heinous of mortal sins.

In Renaissance Europe, this taboo began to be challenged, to the extent that the iconography of man-made flight became synonymous with great cultural and technological rebirth. Enlightenment meant looking toward and venturing into the skies, shaking off earthly bounds, to a totemic degree—Caspar David Friedrich's *Wanderer above a Sea of Fog*, Benjamin Franklin launching kites in lightning storms, the Flammarion engraving with its voyager breaching the firmament to understand the mechanics of the universe. Leonardo da Vinci's obsession with aeronautics began early, long before his prototypes of fantastical flying machines. In his *Codex on the Flight of Birds*, he claimed that his sketches of the flight patterns of birds, particularly the fork-tailed red kite, were predestined, "When I was in my cradle... a kite came to me, and opened my mouth with its tail, and struck me several times with its tail inside my lips."[5]

Faith in mechanical ingenuity and physical courage began to replace that of religious doctrine. Yet superstitions and the idea of heretical trespass of the skies survived into the age of so-called Balloonomania. Having achieved the first-ever hydrogen-fueled balloon flight in 1783, Jacques Charles and the Robert brothers were attacked by the God-fearing, pitchfork-wielding villagers of Gonesse, ten miles northeast of Paris.[6]

It's notable that the first aviator-inventors were polymaths, but they weren't confined to academia. Their studies were essentially live and in the field. Chance and craft played a part. They were open to seemingly unrelated observations and to adapting existing technologies, whether the inspiration the balloonist Montgolfier brothers took from freshly washed linen billowing next to a fire or the Wright brothers, who began their airplane designs out of their family's involvement in the bicycle trade. There was, of course, plenty of trial and error and dead ends—John Stringfellow's aerial steam carriage (Ariel) and the Robert brothers' belief that one could row through the sky with oars, among them.

Other aeronautical lineages developed outside the West, too. In India, for instance, there was the long-standing story of Vimāna, flying palaces in the Vedas, which were claimed in the likes of the *Vaimānika Śāstra* to have been actual aviation vehicles. These were said to have inspired the questionable claims that Shivkar Bāpuji

Talpade flew a mercury-powered bamboo plane over Chowpatty Beach in 1895, which, like the pre–Kitty Hawk claims of flight by the New Zealand farmer Richard Pearse, demonstrate that the entrepreneurial desire to fly, if not ability, was global.

Skyward explorers were brave, by necessity, from the beginning, but they were not suicidal. When the time came to launch the Aérostat Réveillon, King Louis XVI recommended that prisoners be used as expendable test subjects. Instead, a duck, a rooster, and a sheep christened Montauciel ("climb to the sky") were dispatched in the balloon's basket.[7] Humans would soon follow and find that fear was not unfounded. The meteorologist James Glaisher was lucky to survive his record-setting ascent to, and loss of consciousness at, thirty-six thousand feet, the height at which pressurized commercial aircraft now fly. An attempt to surpass this journey ended in the deaths of the aviators Joseph Croce-Spinelli and Théodore Sivel (their tomb in Père Lachaise depicts them holding hands in death). The only survivor, Gaston Tissandier, recounted how they'd behaved irrationally, drunk on hypoxia: "One becomes indifferent, one thinks neither of the perilous situation nor of any danger; one rises and is happy to rise."[8]

New technologies tend to be understood and represented in preexisting ways rather than in and of themselves—*aeronaut*, for example, derived from the ancient Greek for "air-sailor," and *astronaut* from "star-sailor." Given that it was a rare opportunity in a modern industrial world of factory lines and machine guns to exhibit individual bravery, there was a sense that early flying was a chivalric pursuit. These were knights of the air, like "the Flying Baron" Carl Cederström and the "Red Baron" Manfred von Richthofen or daring high-society socialites like Aida de Acosta. This aristocratic tendency existed across the globe. The Japanese view that pilots were the samurai of the skies gave rise to the era of the kamikaze. For a time, flight seemed the preserve of those with money, status, and time on their hands. Yet the extreme danger and the do-it-yourself aspect of early plane design was such that it attracted not only those seeking glory but also those seeking escape from class, sexual, and ethnic confines. The Norwegian aviator Dagny Berger had previously worked as a maid. "Queen Bess" Coleman came from a family of sharecroppers.

The danger *was* real and part of the public appeal, which became fevered. Over the century of Balloonomania, there were many fallen heroes and heroines, and much salacious press. Rozier and Romain were "dashed to pieces."[9] Sophie Blanchard ended up a "shattered corpse" after her balloon was ignited by fireworks and she toppled out onto the rooftops of the rue de Provence.[10] Thirty years after Solomon Andrée's Arctic balloon expedition of 1897 crash-landed in the Arctic wastes, ninety-seven photographs were found on frozen rolls of film, documenting their doomed attempt to return to civilization.

Initially, the combination of photography and man-made flight had been immensely fruitful. It found its apotheosis, and greatest salesman, in the figure of Gaspard-Félix Tournachon, or "Nadar," as he was known. A gifted cartoonist and socialite portraitist, Nadar began capturing Paris from the air in 1858, building his own soon-to-be iconic balloon *The Giant* five years later. Believing the future was in steam-powered proto-helicopters, he formed the Society for the Encouragement of Aerial Locomotion by Means of Machines Heavier Than Air and the journal *Aeronaute*, with his friends the Vicomte de Ponton d'Amecourt and Jules Verne. Set tantalizingly between the fantastical and the feasible, the latter's books would go on to inspire the careers of countless future pilots and successive generations of rocket scientists, by their own admission. There's a case to be made that his *From the Earth to the Moon* led to man reaching the moon, following Nadar's maxim that "all that is not absurd is possible; all that is possible may be accomplished."[11]

Nadar's perilous, self-mythologizing exploits made him a folkloric hero, but he was courting as well as defying death. In the year of his first flight, he was already formulating how useful aerial photography would be in the mapping and the waging of war. He would utilize this synthesis of technologies practically and with great inventiveness when the Prussians laid siege to Paris in 1870, sending communications from the capital to the outside world by balloon and microfilm. Looking back, the photos he took of the city of Paris still retain some degree of the awe they must have once inspired, if not the shock of the new perspectives. With historical knowledge of what was to come, one has much more of a sense of the immense vulnerability that flight had just opened up on this city and all others.

Death was there right from the infancy of the airplane; the first was during a Wright Brothers' flight at Fort Myer when the young lieutenant Thomas Selfridge shattered his skull. Its shadow was always there on long-distance ventures—picture Amy Johnson, Jean Batten, and Amelia Earhart battling through sandstorms or tropical cyclones, with one eye on their fuel or their burning engine or the night waves just below their wings.

Knowing that danger was the draw, impresarios ramped up the risks. Flying circuses and wing walking became part of a decadent age of speakeasies, carnies, and dance marathons. Pilots, like Lindbergh, got their break as barnstormers. Some became celebrities, even household names—the hard-drinking "Bad Boy of the Air" Bert Acosta, the "Flying Schoolgirl" Katherine Stinson, the "Flying Gypsy" Leslie Hamilton.

Sometimes, spectators got what they paid for. "The Man Who Owns the Sky," Lincoln Beachey, smashed into San Francisco Bay while performing stunts in front of a quarter of a million people at the Panama–Pacific International Exposition. "Queen Bess" plummeted from a negligently maintained plane, yet no one was ever charged. Vicious rivalries were not uncommon. When B. H. Delay's plane failed during a loop, an inspection of the wreckage found evidence of sabotage.

Fame offered no protection. The Italian aviatrix Gaby Angelini was swallowed by a dust storm in the Libyan desert. Feng Ru, star of Chinese aviation, crashed into a bamboo grove. Having bailed out, the world-famous Amy Johnson was pulled under, between two ships, in the Thames. Princess Anne of Löwenstein-Wertheim-Freudenberg vanished over the North Atlantic. Jorge Chávez Dartnell's airplane was pulled apart by alpine winds. His last words were reputedly, "Higher. Always higher."[12]

Before flight recorders, aerial disappearances and disasters were often enigmas, inviting speculation. Writers began to fill the vacuum, populating the vast ocean of air above us with creatures, as of the deep. Arthur Conan Doyle's "The Horror of the Heights" proposed malevolent jellyfish-like organisms above the clouds, and Roald Dahl's saboteur *Gremlins* emerged from actual Royal Air Force folklore.

This was the shadow of a golden age of aviator Romanticism, when writer-pilots like Guy Murchie (*Song of the Sky*), Ernest K. Gann (*Fate Is the Hunter*), and especially Antoine de Saint-Exupéry (*Wind, Sand and Stars*, *Southern Mail*, *Night Flight*) wrote of their flights in ecstatic terms: "The magic of the craft has opened for me a world in which I shall confront, within two hours, the black dragons and the crowned crests of a coma of blue lightnings, and when night has fallen I, delivered, shall read my course in the stars."[13] Far from defiling heaven, they had restored its sanctity. Theirs was a literature of awe.

To understand why this period didn't last, consider the nature of awe, which consists of not just wonder but also terror. It was long known that higher ground meant military advantage. And there is no ground higher than the sky. Governments were surprisingly slow in recognizing this. Solomon Andrews impressively demonstrated the potential of his steerable airship, the *Aereon*, to Lincoln and yet was dismissed. Others went further: where amazement wouldn't convince, perhaps fear would. As a warning of what air superiority meant, Lincoln Beachey buzzed the White House, while in Kensington, London, Stanley Spencer rained rubber balls down onto the streets. During World War I, the Italian nationalist Gabriele D'Annunzio dropped hundreds of thousands of leaflets onto the Austrian capital: "We are flying over Vienna; we could drop tons of bombs... PEOPLE OF VIENNA, think of your own fates. Wake up!"[14]

The waking up would be brutal. It would happen all over the world. Although there were many disturbing warnings—Picasso's *Guernica*, H. S. Wong's photograph of a baby screaming in the bombed-out wreckage of Shanghai—it was hard to fully conceptualize until experienced: Barcelona, Abyssinia, Warsaw, Rotterdam, London, Chongqing. "My child's conscience wasn't so much struck by the word *war* as frightened by the word *airplanes*," Dima Sufrankov remembered.[15] Such was the enormity of destruction—Dresden, Tokyo, Hiroshima, Nagasaki—that modern analogies seemed no longer adequate, and the biblical returned, from avenging angels and horsemen of the apocalypse to the raining down of fire and brimstone upon Sodom and Gomorrah. But the prophets who had foretold this were not religious in nature as much as they were simply astute readers of the human soul. It was not God, who is silent, or technology, which is amoral, that caused this, but us. "The machine does not isolate man from the great problems of nature," Saint-Exupéry wrote, "but plunges him more deeply into them."[16]

It's tempting to locate a point at which Romanticism about the air died. One contender would be the futurists, especially their aviator-painter Tullio Crali, who captured the thrilling verve and risk of flight in his kinetic works. Then, abruptly, in his *Nose Dive on the City*, from 1939, we find that elation turn to rage, heroism to cowardice, theatrical play to carnage. The pilot attacks the city, as Stukas would with their terror-inducing Jericho trumpets. The victims are unseen or rendered abstract by distance. The murder is mitigated. In truth, abstraction has existed since we invented weapons that blurred the connection between cause and effect and perspectives that transcended human proportion. I type these words in the Dolphin Tavern, Holborn. There is a charred clock forever stopped at the time of 10:40, the exact moment that a bomb, dropped from a German zeppelin, destroyed the building and its occupants. What was this place to the pilot other than distant geometry?

In the period of relative optimism and temporary stability that followed World War II, it seemed as though the wild beast of the skies that humanity had unleashed might be tamed. Wonder and terror were, it seemed, lost with the domestication of commercial flights. Flying over the Alps, the South China Sea, the Caucasus, Papua New Guinea, I've looked around to see not a soul on the plane looking out the window, and suitably disgusted, I ordered a drink, turned on a screen, and joined them. Meanwhile, aerial warfare was exported elsewhere, in a renewal of the imperial models of previous centuries. Its cruelties—napalm, Agent Orange, daisy cutters, cluster bombs, white phosphorus, thermobaric weapons—tested on other people. There have been numerous incidences in recent years of children in conflict zones being terrified of the blue sky due to invisible drones. "A lot of the kids in this area wake up from sleeping," the thirteen-year-old Yemeni boy Mohammed Tuaiman pointed out, "because of nightmares from [drones] and some now have mental problems. They turned our area into hell and continuous horror, day and night, we even dream of them in our sleep."[17] Tuaiman would be murdered, like his father and brother, by a strike from the sky weeks later. This is the sanitized, remote-control version of aerial war offered by and for the guilt-free operator.

The normalization of the extraordinary, the abstraction of other people and places into mere shapes, endangers us all. This is why it's important to fear, no less than it is important to feel wonder. It keeps us connected to reality and to one another. A pleasant dulling of the senses occurs otherwise, a slow dissipation of existence, consequence, conscience. Sudden surprises might startle us back into life, into the realization that we are aloft in an aluminum tube weighing two hundred tons, flying faster than any Formula 1 race car and higher than Everest. Epiphanies do occur. Imagine being on board the passenger plane that, veering off course, rediscovered the Nazca Lines. Or simply being on a flight that suddenly encounters severe turbulence or a lightning strike. Works of art may have that power on occasion (James Dickey's *Falling,* for example), but it's hard to awaken from the reverie or torpor—the chore of airports, the ticking of the clock, the ease of flying on the safest form of travel. All of which encourages us to forget the fact that this was, for millennia, a utopian dream, the perfecting and corrupting of which has cost countless lives. It is symptomatic of a larger modern malaise: the illusion of weightlessness, the ignorance of what and who got us here and at what cost, the ingenuity and hazards involved, the fuel and the waste, the carbon. Outside our shuttered airplane portholes are the vast glories of the heavens and below our seats, the abyss, and how vital it is, for the sake of our very souls, to cling to the knowledge of both.

notes

1. London Lancet Commission, "The Influence of Railway Travelling upon Public Health," *The Lancet* 79, no. 2002 (January 11, 1862), 107–110.
2. Charles Zika, *The Appearance of Witchcraft: Print and Visual Culture in Sixteenth-Century Europe* (New York: Routledge, 2009), 173.
3. John Milton, *Paradise Lost* (n.p.: Homer Baxter Sprague, 1879), 13.
4. Howard Williams, *The Superstitions of Witchcraft* (London: Longman, Green, Longman, Roberts, & Green, 1865), 82.
5. Walter Isaacson, *Leonardo da Vinci* (New York: Simon & Schuster, 2017), 19.
6. Richard Hallion, *Taking Flight: Inventing the Aerial Age through the First World War* (New York: Oxford University Press, 2013), 51.
7. Charles Coulston Gillispie, *The Montgolfier Brothers and the Invention of Aviation* (Princeton, NJ: Princeton University Press, 2014), 47.
8. Hallion, *Taking Flight*, 77.
9. *European Magazine and London Review*, no. 42 (Philological Society of London, 1802), 30.
10. Deborah Noyes, *Lady Icarus: Balloonomania and the Brief, Bold Life of Sophie Blanchard* (New York: Random House Studio, 2022), 130.
11. John Wise, *Through the Air* (Philadelphia, PA: To-day Publishing, 1873), 145.
12. Willie Hiatt, *The Rarified Air of the Modern: Airplanes and Technological Modernity in the Andes* (New York: Oxford University Press, 2016), 37.
13. Antoine de Saint-Exupéry, *Airman's Odyssey* (New York: Reynal & Hitchcock, 1942), 12.
14. Edgar Charles Middleton, *The Great War in the Air* (London: Waverley Book Co., 1920), 4:189.
15. Svetlana Alexievich, *Last Witnesses* (New York: Penguin Modern Classics, 2020), 153.
16. Saint-Exupéry, *Airman's Odyssey*, 40.
17. Chavala Madlena, Hannah Patchett, and Adel Shamsan, "We Dream about Drones, Said 13-Year-Old Yemeni before His Death in a CIA Strike," *The Guardian*, February 10, 2015, https://www.theguardian.com/world/2015/feb/10/drones-dream-yemeni-teenager-mohammed-tuaiman-death-cia-strike.

Castles in the Air

Nicholas Triolo

Huffing 90 percent pure oxygen through a blue-tubed nasal cannula, I ask the air saloon associate, Cameron ("Like Diaz!"), if I'm doing this right, to which she nods and continues to massage my neck with an arachnid-shaped roller while two electric pads stuck to the small of my back provide shocks at half-second intervals just as I try and make sense of a (fake) sky painted in cumulus above two (real) lovers tongue kissing on a (fake) gondola boat rowed by a (real) retiree in a striped shirt along a (fake) Venetian canal, all set to the tune of (real) opera piped through (fake) rocks that double as speakers—and somehow, in some strange way, everything I know about the (real) world starts to make a lot more sense.

In July 1776, two years after oxygen was officially "discovered," Thomas Henry, a renowned surgeon and chemist from Manchester, England, was the first on record in the *London Review of English and Foreign Literature* to imagine air as a recreational commodity: "This dephlogisticated air will soon come to be sold like ice at the confectioners, or to be as fashionable as French wine at the fashionable taverns." He went on to suggest that air-for-sale might soon be available in apothecaries, in labeled vials. He hoped the trend would "convert their natural privilege," that clean air would be something to leverage for the luxury class, something to be controlled, owned, made into currency.[1]

Imagine if a mescal bar downed a fistful of MDMA and fucked a chemistry set. That is what an "air saloon" in Las Vegas looks like.

In *The Man Who Sold Air in the Holy Land*, the Israeli novelist Omer Friedlander tells the story of an old man, Simcha, who sells bottled air to tourists in Jerusalem. Simcha's previous scam included filling vials with tap water, adding a pinch of salt, and calling it sacred water from the Dead Sea. This proved too complicated, but air?

"Air was everywhere," writes Friedlander. "He didn't need anyone's permission to bottle it, and most important, it was free." Simcha takes care to make the air appear fancy, "like the Mediterranean cabernet sauvignon of the Judean Valley." He even designs a label for his new product with a Greek god blowing gusts of wind. His daughter, Lali, carefully prepares each bottle, organizing them all on a rickety blue trolley before heading out on Fridays looking for customers.

One day, Lali and Simcha spot a rich American couple window-shopping and create a setup where Lali will act as a stranger and fake interest in Simcha's bottling trick.

"Air?" says Lali. "Why would anyone buy air?"

"This isn't just any air," says Simcha. "It's very special. I'll tell you a secret."

He whispers gibberish into her ear, to which Lali exclaims, "I want ten of them!"

She has no money and even goes so far as to propose selling her one-eyed cat for the air, to entice the onlooking tourists. Lali turns to the couple, asking if they'd buy her cat so she could afford the purchase.

"What's in them anyway?" the tourists ask Simcha, now curious.

"Oh, in these?" Simcha says. "Nothing. Only air."

The couple deliberates; the tactic works. With their meager earnings, Simcha and Lali enjoy popsicles and falafel. They visit an abandoned SeaWorld-type attraction called the Dolphinarium and sit and imagine dolphins jumping through the air.

But Simcha grows increasingly poorer every year. The old man sits on his stoop and calls into the streets for future clientele, but people walk by saying: "We're breathing it for free, you donkey!" The story circles in vacancy: the bottling of nothing, the abandoned aquarium, empty promises and empty pockets. Near the end, the father dresses up in his fanciest suit and takes Lali to eat at a hotel buffet, where they gorge on every imaginable confection until it comes time to pay. Simcha tells the waiter to bill their room—a room number he makes up—and the waiter refuses. The old man has no money and loses his cool until the daughter pulls from her tin box the last remaining profits from conning people into buying air.

Eventually, Simcha gets evicted. His bottles of air disappear. Lali asks her father, now homeless, where he'll sleep. "Anywhere," said Simcha. "The Hilton, the Dolphinarium, a castle in the air." *A castle in the air*: fortunes built without foundation.

The story ends with Simcha imagining all the ways he'll make that next profit, from raising chickens and painting their eggs gold to training parrots to talk as if they could transmit voices from the dead. Lali loses faith: "He kept talking and talking, even though he knew that she'd stopped listening anyway—soon she would stop believing him, just as he had stopped believing in himself."[2]

There are thirteen registered oxygen bars, or air saloons, along the Las Vegas Strip, but the one I visit sits chambered deep within the Venetian Resort. Each of the kiosk's ten hookups offers proprietary air from distillation machines that bubble blue liquid, appearing to be Windex. Cameron attempts to upsell me several massage tools, distractions from her dubious role as air associate. But she wouldn't use the word *dubious*, and neither would I, because after purchasing my air treatment for $1 per minute, I realized this might actually be the single-clearest expression of something I've been trying to further understand: the transmutation from public resource to private commodity—the sale of the commons. And just as the oxygen begins tickling my nose, just as Cameron's Pepto pink fingernails adjust my nose piece, I wonder, *What could be a clearer expression of commons-turned-commodity than selling... air?*

In Jules Verne's 1870 novel *Around the Moon*, three men launch a rocket and orbit the moon before returning to Earth and splashing down in the ocean. At one point in the journey, they experience an oxygen leak in the astronaut's living quarters, which saturates their capsule, likely to kill them by combustion if they don't do something about it. The three become excessively high until the capsule regains equilibrium, and they fall into a hungover torpor, "obliged to sleep themselves sober over the oxygen as a drunk does over wine." Instead of being worried, though, the astronauts quite enjoy the oxygen's trance; the leak broke up the monotony of their moonshot. Michel, one of the Frenchmen, imagines a future in which "a curious establishment might be founded with rooms of oxygen, where people whose system is weakened could for a few hours live a more active life. Fancy parties where the room was saturated with this heroic fluid, theaters where it should be kept at high pressure; what passion in the souls of the actors and spectators! What fire, what enthusiasm!"[3]

The proliferation of the modern air saloon began in the 1980s in Japan. Called *sanso* bars (Japanese for "oxygen"), the first of its kind debuted in 1988, in Tokyo's major department store, Takashimaya.

A network of the air stations emerged not only as an offering for the aristocracy but also as a response to growing concerns of air pollution, given the country's rapid industrialization starting in the 1870s and lasting through the twentieth century. A *New York Times* article in 1964 referenced copper roofs in Tokyo turning black from high sulfuric acid in the air and claimed that the "sky over Tokyo has disappeared."[4]

The air saloon trend grew internationally, starting in 1996 with a recreational oxygen bar in Toronto, then spreading to the Rocky Mountains, where chic mountain kiosks were touted to help flatlanders acclimate, and later, finding its way here to Las Vegas, with stations installed for hungover tourists wanting to recover quickly and get back in the game. It's not uncommon now for oxygen bars to offer a full menu of flavored air, including mandarin orange, green apple, and lavender. And in Japan, where this whole trend began, in 2007 the company Air Press launched twenty locations exclusively for canine oxygen bars, where pets are put into capsules infused with 100 percent oxygen for sessions that "offer the same aerobic benefit as two hours of exercise."[5]

"Who's the biggest celebrity you've seen stop by your oxygen bar?" I ask.

"Well, we've seen Pauly D, so that was cool," says Cameron.

"Who's Pauly D?"

"You don't know who Pauly D is?"

"No."

"*Jersey Shore*? Oh my *gawd*."

"OK, so why are my oxygen tubes connected to these bubbly blue vials of water?"

"The air gets filtered through that water so it doesn't dry your nose out."

"Oh."

"At end of the bar, we have three machines called oxygen concentrators. Each pulls in ambient air, compresses it, and filters out nitrogen so that all you're left with is pure oxygen. But pure oxygen can be *dry*. And the color? I mean it's scientifically proven that blue makes people *chill*. That's why our whole thing is blue."

"You must see some pretty crazy stuff, managing an air saloon in Vegas and all."

"*Literally* all the time. In our Stratosphere location we have these lay-down massage chairs to start everyone out before moving them to bar stools. One guy fell asleep for hours. Woke up in full panic mode. 'Where am I? Who am I?' he said. We have super drunk people falling out of chairs and passing out. *Not because of the oxygen.* I once had a customer who had her drink roofied."

"Was she OK?"

"Fainted. Had to call EMTs. *Not because of the oxygen.* I get people walking by all the time saying, 'This is dumb, we can breathe for free!' And I say, like, all you're breathing out there are cigarette smoke and farts."

"Cigarette smoke and farts?"

"Cigarette smoke and farts."

"Interesting."

"Look, ambient air is *only* 20 percent oxygen. If this were a scam, they wouldn't use oxygen in hospitals, right?"

"Right."

"Oh, and get this: the other day a man walked by and I said to him, 'Hey, you should come try this,' and he was like, 'Hey, I'm a respiratory therapist and you're trying to sell water to a fish right now,' and I said, 'Hey, fish would *die* without water.'"

Both of us in unison: "Burn!"

Successfully transitioning any public good to a privatized commodity requires a three-step program. The first step: radical objectification. Water, air, and earth must have all animacy, all sentience, all divinity revoked to be reimagined as something vacant and inert, something capable of being fractionalized and controlled. The second requirement follows close behind: the installation of fear. Fear of taint. Fear of lost purity. Fear of pollution and scarcity. This sets the stage for the final step: privatize and appeal to acts of self-care, class identity, and eco-consciousness.

Nestlé—which in 2022 raked in more than $10 billion in profits—is known for systematically acquiring headwaters and springs for its water-bottling empire. The company often appropriates sacred waters from Indigenous communities and deploys tactical fear campaigns: that municipal water, an enemy of purity, isn't to be trusted; that citizens must instead trust in bottled products to ensure their family's welfare. Nestlé's current water campaign is appropriately named "Pure Life," and it encourages consumers with the leading slogan: "Drink Better, Live Better."[6]

Invest in bottled water or containers of recreational oxygen, and you're investing in your health—or better yet, the health of the planet. Take a deep inhale. Taste the minerals. Smell youth. Smartwater. Karma Wellness Water. Cans of air are now available online for purchase, direct from Zermatt, Switzerland.[7] *Treat yourself.*

In the 2012 animated film adaptation of Dr. Seuss's 1971 book *The Lorax*, O'Hare Air is a company that sells bottled air to a town made entirely of plastic called Thneedville, run by its mayor, Aloysius O'Hare. In an early scene, two executives pitch the idea of selling bottled air to citizens. "You really think people are stupid enough to buy this?" asks O'Hare. After mulling over the idea, O'Hare exclaims, "In other words, the more smog in the sky, *the more people will buy!*"

Instead of relying on oxygen supplied freely by trees, which are being logged out of existence beyond the city's walls, residents of

Thneedville are instead sold air by O'Hare. First, trees are othered and objectified, reconstituted as enemy. Next comes the fear: their sticky sap and stinging bees are marketed as a nuisance, a threat to be removed. And finally come appeals to lifestyle and conviviality. In one O'Hare Air commercial, two young men sit bored on a rooftop until one cracks open a bottle of O'Hare Air. Immediately, friends barge onto the roof with umbrellas and six-packs of air and start dancing and chugging bottles. An attractive blonde woman huffs a bottle and winks at the camera. The men clink their bottles.

"O'Hare Purified Air, Freshness to Go," says the slogan. "Please Breathe Responsibly."

Back at the Vegas air saloon, Cameron adjusts my nose piece as she complains about how often people critique selling air. Many think it's a scam, she says, even though everyone still walks around with bottled water. "I'm like: *Who's being scammed now, bitch?*"

Indeed, total US bottled water sales in 2023 are projected to be $94 billion, and that number will likely grow annually by 6 percent.[8] Such profits follow the three-step playbook: objectify, fear, develop market distinction. As seen in cases like Flint, Michigan, some municipal water sources are indeed harmful, but those instances are rare. In the case of air as commodity, though, public air can be harmful. Air pollution is an increasing problem. Lung cancer and asthma are linked to increasing wildfire seasons and lingering smoke. The pandemic especially brought the value of air to the forefront. May 2021, for example, saw the single-highest rates of COVID-19 cases and deaths in India, requiring an average of two million canisters of medical-grade oxygen per day. This immediately spawned a black market for air canisters, which were hoarded and sold at astronomical prices. Normally, a conventional canister of oxygen costs anywhere from $81 to $135, but during the worst of the crisis, canisters were sold for over $1,300.[9] This turned air into a form of currency, used to decide who survived and who died.

Whether air or water, the impulse to turn a public health concern into a market opportunity, to repackage precarity as a doorway to wellness, is particular to the times in which we live. "We've gotten here, step by step, down a dangerous road of converting a public resource into a private commodity," said Peter Gleick, scientist and cofounder of the Pacific Institute, which is focused on global water health. "Water utilities don't have advertising budgets; private companies do."[10]

Clear Lounge, the world's first underwater oxygen bar, caters to Carnival Cruises in Cozumel, Mexico. For twenty minutes you are submerged in a full-glass water tank with an oversized scuba headset and piped aromatherapy oxygen while people spectate. During the session, oxygen bar patrons play underwater Jenga, shoot each other with bubble guns, and draw on waterproof canvases. "When you get out of Clear Lounge," says the website, "don't forget to try the delicious oxygen-infused smoothies or purchase canned oxygen to take home with you!" Clear Lounge has recently opened a second location in Kuwait.

In many ways we *do* pay for air, if not as recreation, then as utility. It's unlikely for Fiji air to be available at the corner market any time soon, but we can still purchase air in medical emergencies, or as HEPA filtration systems, or as air conditioning. Is *conditioning* air some indirect form of commodification? After all, US homeowners spend nearly $30 billion annually on air conditioning.[11] The definition of air as commodity grows complicated and diffuse, but maybe that's the point—air is something of which we are part, with which we are in binding relationship. It does not exist *out there.*

My time at the air saloon has come to a close. I arrived with the assumption that this experience would be the waxiest fruiting body of late capitalism I could possibly pluck, but I now think that I missed the point, which is that the commons have been sold back to us for hundreds of years, abstracted and repackaged in ways many of us cannot see, or are simply unwilling to. Here in the Venetian, I start to wonder: *Am I the commons, too?*

My visit to the oxygen bar proved highly unusual, but it brought me to the steaming lip of the Anthropocene's cratered center. It laid bare the objectification of elements as ubiquitous as air and water, the turning of abundance into scarcity, the abstraction that maintains the separation between humans and the air we breathe, what the naturalist philosopher David Abram called the "commonwealth of breath," a shared atmosphere filled with language and exchange.[12] And here, in this commonwealth, I think of Simcha, the poor bastard who tried selling air in the Holy Land. Simcha, the con man. Simcha, the broken father, alone, trying to take care of his daughter, so full of promise, so full of life. Simcha, homeless and bailed out by his own child, exiled and alienated yet holding firm to his delusion of eventual bounty, building his castles in the air, a house without foundation, a racket to which he held firm in the belief that air was ever something separate, ever something outside our bodies. Simcha, in the end, learns that he was bottling and selling himself the whole time.

notes

1. Thomas Henry F.R.S., "Essays Physical and Chemical by M. Lavoisier—Translated from the French, with Notes, and an Appendix, by Thomas Henry," note from W. Kenrick, *London Review of English and Foreign Literature* (T. Evans, Pater-Noster-Row, 1776), 4:214.
2. Omer Friedlander, *The Man Who Sold Air in the Holy Land: Stories* (New York: Random House, 2022).
3. Jules Verne, *Autour de la Lune* (Pierre-Jules Hetzel, 1870).
4. "Tokyo Troubled by Air Pollution; Emphasis in Smog Control Is Still on Persuasion," *New York Times*, December 26, 1964, 4, https://tinyurl.com/2vyhs2ss.

5. Akiko Fujita, "Hyperbaric Chambers Breathe New Life into Pampered Pets," *ABC News*, April 8, 2013, https://abcnews.go.com/blogs/headlines/2013/04/hyperbaric-chambers-breathe-new-life-into-pampered-pets.

6. "Nestlé Reports Full-Year Results for 2022," Nestlé Global, February 16, 2023, https://www.nestle.com/media/pressreleases/allpressreleases/full-year-results-2022.

7. See the website of Mountain Air Zermatt—Pure Air from the Heart of the Swiss Alps, at https://www.mountainair-info.ch/shop/.

8. "Bottled Water—United States | Statista Market Forecast," Statista, https://www.statista.com/outlook/cmo/non-alcoholic-drinks/bottled-water/united-states.

9. Shamani Joshi, "India's Black Market for Oxygen Is Booming: Only the Ultra-Rich Can Afford It," *Vice News*, May 4, 2021, https://www.vice.com/en/article/7kv95q/india-black-market-oxygen-only-the-rich-can-survive-covid.

10. Emily Stewart, "Bottled Water, Boxed Water, and the Scam of Selling a Natural Resource," *Vox*, November 3, 2022, https://www.vox.com/the-goods/23433132/best-bottled-water-is-tap-environment-health.

11. "Air Conditioning," Energy.gov, https://www.energy.gov/energysaver/air-conditioning.

12. David Abram, "The Commonwealth of Breath: Climate and Consciousness in a More-Than-Human World," Center for the Study of World Religions, April 9, 2019, https://cswr.hds.harvard.edu/news/2021/05/10/video-commonwealth-breath-climate-and-consciousness-more-human-world.

On Blessings & Want

Felicia Zamora

A Cumulus cloud streaks the belly of horizon, reminds me of the illusion of stillness; both inside & outside the flesh. How I used to long to be still—now, I am learning to long to understand the lack of stillness. A bee doesn't think of its buzz; nor does a bee flap its wings. Only nectar & hover & pollination & honey & hive. Innate buzz cradled intricately inside the flight muscles pulling on the springy thorax wall, in & out, to create a *ping ping ping* sound. A bee's muscles contract multiple times from a single nerve impulse & combined actions make the bee beat each wing—each wing in cycle 230 times per second. Each minute a 13,800-wing-beat blessing. Each buzz brought by Wind to an ear canal, another blessing. I search for blessings now. I find myself flapping; anatomically & cognitively going nowhere; I still cannot comprehend some of my own patterns; how beating in my organs not only doesn't cease, but sometimes I feel wounds from my own imaginary fists punching behind my rib cage, behind my ocular bones. I want to heal in a world where carbon emissions don't eat the world first; where greenhouse gases stop trapping heat; where the pumpkins rotting in the dump don't turn to methane; all the methane building & those ghostly fists pummel & pummel; where I don't first worry about what the clouds bring, before worrying about the clouds themselves. A wickedness to want wanting. So I learn from the clouds. Look for blessings before the frost & within each crystalline formation. I open to stillnessless in all my biological humming. Bless my openings, my permeable epidermis. Bless my surfaces seen & unseen. Bless my jaw in boney hingeness. Bless my mutation of mutability, cloud-esque, from Stratocumulus to Nimbostratus to Cirrocumulus—let me paint the sky. Bless the air that moves through me. Bless the bees. O please let the bees become un-endangered. Bless floating water. In all my formations, I bow to Mother Wind, in her gentle embrace, terrible embrace; behind my breastbone a *thud, thud.* Bless me Mother, I am your hapless daughter who forgets you, then falls to my knees, myocardium gasping, *Shape me.*

We Will Know Ourselves Beloved

Sara Beck

Just before the fall, I can see it all: the identical roofs of Bojrab Drive, the suburban-straight lines of utility poles, and, just beyond the tree line, the corn rows of the last remaining farm. I am nine years old, and I have never climbed higher. Not when I had shimmied up our basketball hoop to distract my older brothers during their neighborhood game. Not when I had helped my dad clean out clumps of leaves from the roof gutters. This is different. Blue-sky high. Bird-wing high. Heaven-bound high.

Perched on that ladderlike pine branch, I lean outward, a wisp of a thing hanging in the air, until suddenly the branch shivers beneath me, my foot slips, and I am falling. Slow, slow, as if in a dream. Pine needles whisper past my cheek. Branches ease my speed. And there, just to my left, I am sure I see an angel. Not just any angel... *my* angel. With the sudden insight of a spiritual revelation, I understand that while she is invisible most of the time, she is always here with me whether I see her or not. Holding me, guiding me, directing and redirecting me. Her wings help me now as I float downward in a lazy, feathery fall.

This isn't bad, I think, but then it changes. Just like that, I land hard. A packed-dirt, root-ground hard. My spiritual X-ray vision and angel-feather air have abandoned me, and along with them, my breath. I gulp, my mouth twists open, I struggle, but nothing happens. As if suctioned out by a vacuum, my breath has left my body, and I am buried, a dark and heavy lump.

From far away, my sister's voice breezes down.

Please don't die.

The air carries her voice. It falls to me, weighted with longing. *Please. Please don't die.*

Once upon a time, the Upanishads report, the sense organs of the body fought with one another. Tongue and eyes and ears and mind and breath, each claimed to be the best.[1] They jostled and postured until the creator himself advised an experiment, promising that the greatest would be revealed by whose absence was felt the strongest.

They settled in for the investigation, with the body serving as journalist. One by one, tongue, eyes, ears, and mind left for a year at a time, and upon their return, the body reported slight variations of impact. The absences were challenging, but they were manageable.

Finally, it came down to breath. The body braced itself while the other sense organs watched, expecting a similar yearlong experiment, but as soon as breath began to depart, the body suffered. It gasped, weakened, and felt itself flattened. A mile underground, the body cratered into a grave.

Beloved! the senses begged, *Please come back! Without you we are nothing!*

Breath. Vehicle of air. That most elusive of elements. A presence invisible, understood only through evidence of its visitation. A curtain rippling. A mirror fogged. The skin blasted hot or breezed cold. The belly, rising, falling. This air, full of gas and aerosol and particulate matter. Measurable in part, yet remaining a mystery. Life itself, whose greatness is known through absence.

As long as we breathe air we are in life and life is in us.

When I was a kid, I never thought about my breath unless I was being tickled by an aunt or smothered by a pile of siblings. *Stop! Stop!* I would shout. *I can't breathe!*

Or I woke up gasping in the night to find my sister plugging my nose, her exasperated attempt to stop me from snoring in our shared bedroom.

Or I finished the four-hundred-meter dash, and hanging over my legs, lungs burning, brain dizzy, I'd choke out the words *Let me catch my breath.*

Then I discovered yoga. I was twenty years old, and I felt ancient. My body was a wreck from a decade of competitive high jumping, which had landed me on a Division I college track team until chronic back pain pushed me out. Nothing seemed to help. I endured ice baths, periods of forced rest, physical therapy, bone scans, MRI imaging, chiropractic adjustments, and enough ibuprofen to make my stomach bleed. I submitted to prayer ceremonies and received the laying on of hands. One doctor told me I needed to accept a life of pain. Another said it was all in my head. Finally, I stopped competing. I discovered massage and modern dance, and although both seemed to bring relief and mental joy, the specter of pain remained, an ever-present ache in my lower left side that zinged up and down my leg until, once a year, it froze itself into a twist in my spine. I was bedridden for days.

During this period, my boyfriend, newly returned to Indiana from California, rolled out a yoga mat. *Just give it a try*, he said, sticking a cassette tape of the Ashtanga Yoga Primary Series into his stereo. The Primary Series is a precise sequence of physical yoga postures, linked together and done in a specific rhythm and rate kept steady by a breath count. The cassette-tape teacher instructed us to begin something called *ujjayi* breathing, but as far as I could hear from the stereo speakers, she was imitating Darth Vader or the heavy nighttime sighs of my great-grandmother. Each pose was held for a minimum of five breaths, and the only instruction the tape gave was when to breathe rather than how to breathe. With some bewildering Sanskrit words thrown in, it went something like: *Inhale, lift your arms. Exhale, fold forward. Inhale, gaze forward, exhale fold in. Inhale, exhale, one. Inhale, exhale, two.* And so on. I fumbled through it, glancing often at my boyfriend for help. I was surprised by the muscle-burn challenge of holding simple shapes and delighted by the joy I felt when I got to lay down at the end of the class and *release the effort in the breath*. I felt strong and open, intrigued and inspired to do it again.

Over the following two years, I moved from cassettes to a book with pictures to a class in a local library. I felt my body changing—healing—as it settled into a steadier alignment, less susceptible to sudden shifts of movement and mood. I learned a lot about anatomy, but no instruction helped me understand the breath until I found a local yoga studio and enrolled in Ashtanga for Beginners. The teacher, a petite wiry man in his early thirties, began earnestly. *All breaths*, he said, *are not created equal. The yogis knew this before anyone.*

He explained that to cultivate the full extent of our human, divinely sparked vitality and to heal ourselves from disease, stress, and conflict in the body, we had to learn to use the breath as a tool. The *ujjayi* breath, translated from the Sanskrit as "victorious breath," was just one of many breath practices that those ancient practitioners used to regulate the nervous system and direct the mind.

Without this breath, the teacher said, *the movement practice we do is just a basic stretching routine.* He yawned, as if in boredom. *Nothing special, no transformation.*

A more-than-stretching total transformation? Yes, please!

I soon learned that these breathing practices are called pranayama. *Prana* translates as "life force" or "vital, primary essence." The first breath of God in us. Invisible presence of life itself. *Yama* translates as "restraint," and so pranayama has come to mean breath control, extension of *prana*, the activation of and tuning into the subtle body. Historically taught as a relatively advanced practice to be done only after the body was cleared, strengthened, and expanded through the physical postures of asanas, pranayama invites us to consciously relate to that which is life within us—that is to say, the breath that carries life—so that our primary, vital essence can be fully known, developed, and directed.

Just try it, my first teacher said, echoing my boyfriend. *You'll see.*

He closed his mouth and exaggerated the volume of his breath, creating a tidal quality to the flow of air moving in and out of his body.

Ujjayi *sounds a little like the ocean washing in and the ocean washing out*, he explained.

Following his instruction, I broke down the practice into manageable parts. Breathing in and out through my nose, I made a gentle *haaaa* sound, imagining that I was fogging a mirror. This helped my throat to constrict. Like closing off the flow of water on a garden hose, this constriction allowed me to control the volume and rate of my breath and then to guide it. It would also, the teacher promised, help me to rest my mind, to build focus, concentration, and ease. But instead of calm water washing the back of my throat, my breath bumped in and out like wild bursts from a water gun. I sounded more like myself as a snoring child than a gentle ocean wave.

Within minutes, however, I experienced the sensation of something gentle and smooth: a cooling stream of air down the back of my throat as I inhaled, a warming flow of air expanding outward as I exhaled. Five breaths, ten breaths, several minutes later, I could feel how this intentional breath gathered and held the wayward contents of my mind and drew them inward into an orderly, quiet focus. And I felt something under the effort. A pulse. A vibration. Aliveness itself.

Soon, I began attending class several times a week. I loved walking into the studio, crossing that threshold from hallway into practice space and feeling a palpable difference. The dim light, the candle glow, the hot air, sweet with the scent of sage, sharp with the sweat of other bodies, and thick with the residue of exertion and rest, effort, and ease. I loved how the teacher connected each physical instruction with the action of the breath. I loved how each inhalation supported extension, lift, and a creation of space in my body, and how each exhalation supported release, surrender, and a settling into my skeletal structure. Always, always, the breath was consciously invited to be in service to the body and her movements. Like the current of air filling a boat's sail and determining its direction, so the breath's strength, length, and tempo steered the movement of bones, the range and resilience of soft tissue, and the circulation of blood.

The shape of my body changed. My back pain all but vanished. My mind felt steadier. No longer experiencing myself as ancient or broken, folded in on myself like an accordion, I experienced myself tall, clear, and radiant. A temple space lit up from the inside out.

Since then, for the past twenty-five years, I have practiced many breath styles and forms of yoga. I have chanted and hummed and sung in devotion and in service to my nervous system. I've held yoga poses while crystal bowls, tuning forks, and Tibetan gongs sent waves of sound through my cells. I have studied Ashtanga, Iyengar, Kundalini, and Tantra yoga. I have become a certified yoga teacher, a reiki energy practitioner, and a breathwork coach. I've studied the physiology of the breath and its journey into the nasal cavities, the lungs, and the alveoli. I've loved on the muscle of the diaphragm and poeticized the beauty of its contraction and expansion. I've used sacred plant medicine, visualization, meditation, sensory deprivation, prayer practice, and mantra alongside and with conscious breath. I have done all of this so that I can know—in a bone-deep, muscle-full kind of way—this ever-elusive element of air. Which is to say, I've done this to become intimate with life itself so that when I come to die, I will not find out that I have not lived.

If we drank our breath like water, each gulp would equal about two cups. Elixir of life—78 percent nitrogen, 21 percent oxygen, and 1 percent assorted gases like carbon dioxide, neon, and hydrogen—this air travels into the lungs, where the delicate soap bubbles of the alveoli facilitate absorption into the bloodstream. There are also aerosols, matter that is mostly unseen to the eye. Filmy dust of earth. Black ash of volcanoes. Dewy salt from sea. Microscopic sludge from landfills, animal and human waste, and the by-products of industry. And of course, also contained in this elixir of life are airborne germs and floating bacteria. All of this rides into our bodies on the vehicle of the breath.

Less measurable, the air also holds things felt but not validated by measurement or scale. The weight of a mood. The lightness of a smile. The heaviness of grief. The charge of shock. The warmth of laughter. The chilliness of fear. Finer than gas but just as present, we inhale the lives and deaths of our sister trees and our brother plants. Within the bodies of our lungs circulate the health or disease of all that is living and has ever lived. We inhale the suffering and the joy of our ancestors. We swim in their survival strategies, their fears, and their victories. There is no avoiding it: the air contains the past.

It also creates the future. The contents of our air—carrying the health of our earth, all of its inhabitants, and the well-tended quality of our breath—will become the contents of our children's breath. What do we want their breath to hold?

I was adamant that my son be born into water. I wanted his transition from fish-creature, breathing dark water, to human-creature, breathing bright air, to be softer and less jarring. It felt like a sacred duty to honor the beginning of his life and to cushion his becoming. I imagined the birth tub serving as a symbol of this holy ritual. A natural baptismal font of sorts, it would be something earthy, made from wood and lit by candles, and surrounded by air made fresh with sunlight and sage.

What we had instead was a bright-blue inflatable swimming pool, the kind you see punctured and deflated, sagging against the side of a house at the end of the summer season. Squeezed into our five-hundred-square-foot apartment, it barely fit between my bookshelf and my yellow sofa. I was hardly in it anyway, laboring instead on the couch for several hours, rocking on hands and knees with my head sagging on the arm rest. My support team was always near—my sister and my son's dad, my midwife and her apprentice—and I remember how a hand would sometimes lift my forehead to place an oxygen mask over my face, how I would hungrily sip the air. Instantly, a spark of vital energy. I didn't know why I needed it, but I heard my midwife say something about an *erratic heart rate* and *the baby's distress* and *we must keep her breathing*.

My midwife pressed her forehead against my own. *Breathe with me*, she ordered. *Low. Keep it low.*

By then I had been a yoga teacher for years. I knew all about the relationship between humming low and becoming calm. I knew that my nervous system liked a deep, throat-thrumming sound, which vibrated my vagus nerve and told my brain that I was safe. I knew that I needed to feel safe if I wanted my body to open.

And yet. With pelvis-quaking contractions and bone-shattering pain, I felt only fear. Everything in me wanted to harden against that sensation rather than relax into it.

Life cannot come in or out with tightness, my midwife said firmly. *Breath must be open, low and slow.*

The sound she made was as old as a lullaby. The *om* at the beginning of a yoga class. The chanting liturgy from my childhood church. The sweet moan of pleasure. The throaty groan of relief.

I remembered those first yoga classes, the ocean breath that supported me during challenging postures. In headstand, my arms shook, but I remained steady with ocean breath. In flying crow, my shoulders burned, but I remained steady with ocean breath. In full-wheel backbend, my inner thighs quaked, but I remained steady with ocean breath. All along, I now realized, I was building a unified resilience in body and mind so that I could endure the terrible and beautiful experience that is giving birth, that is allowing the vital essence of life to grow in and pass out of my body.

I knew how to do this. Low and slow, controlled and clear, the breath was my guide into how to open to life, which is sensation, which is pulsation, which is blood moving, which is the pelvic bone widening, which is soft tissue spreading. My body exerted and my mind remained ever-present.

Soon, it was time to push, and my midwife ushered me into the pool. The morning light of late summer made that blue plastic shine, and it was as sacred a site as any Jordan River. As I pushed, my midwife ripped off her pants and jumped into the pool, ministering to me in all that water, and the whole room shouted, all of us exhaling our hope and desire and love for this soon-to-be-born life. A few pushes later, I screamed his body out of my body, and for a split second, he drifted. Down, down toward the bottom of the pool, he drifted, and I saw his mouth open. Then my midwife whisked him into my arms, this dark-haired treasure, and the awe of it all took away my breath like the force of a fall, a fall into a new reality, and all of our breaths were gone.

The room was silent, five bodies tense and alert, listening for the sound of this new body, a sound that was not coming. The air became hot and oppressive with the hope and exertion of the moment. We could not breathe—we *would* not—until this body breathed. He remained still. How many seconds passed? How many lifetimes? His forehead crinkled with effort, lips pressed tight, eyelids clenched into a thin line. Did he feel what the body in the Upanishads felt?

Breathe, breathe, I urged. *Please. Please don't die.*

Suddenly, my midwife snatched him from me, and she covered his mouth with hers. One exhale was all that it took, and he sputtered like a motorcycle engine coming to life. Limbs flailing, he gulped and then screamed. Sharp. Piercing. Perfect. We cheered and we laughed, and the air in the room moved, fresh with new life.

In 2013, researchers in Switzerland discovered that each human being has a unique breathprint. The exhalation serving as signature.

A tool of revelation and transformation, the breath is our connection to all that is unseen within the structures of the body—windpipe and lung tissue, umbrella of diaphragm and journey of nerve—and all that is unseen within the energetic structures of the universe. Gas made visible. God made visible. The past imprinted. The present contacted. The future changed.

My sister is older than me but barely. More like my twin and my opposite. The breath in to my breath out; the steady, clear action to my fiery flights of fancy. She is my first memory. I see her hands clearly. They are holding mine as we spin, surrounded by a room of blue. I always thought I was remembering a dream until my mom told me of the empty dining room where we played when I was a toddler. *We had no furniture,* she remembers. *Just this ocean of blue shag carpet and blue wallpaper.*

Like the sky when I fell. The color of fresh air and possibility.

Please don't die.

My sister exhales her words into the darkness of my casket space. They enter my nostrils and travel into the bodies of my hungry lungs, and just like that, I receive her breath of life. My eyes gasp themselves open. Her tear-wet face swims above mine.

She loves me, I think in wonder.

Smiling, I sigh, and it filters up and up, the sigh absorbing into all that air. The angels wing my breathprint away, offering it elsewhere, to some other body. Falling or fallen, flying and free.

notes

1. *Chandogya Upanishad* 5.1.6–12, from *Chandogya Upanishad*, with the commentary of Sankaracarya, trans. Swami Gambhirananda (Calcutta: Advaita Ashrama, 1983).

Airspace

Michele Wick

I n art, negative space is the seemingly vacant zone surrounding the subject, a place for the eye, ear, or mind to rest. Airspace. Though considered lesser by some, for many it is the heart of form and possibility.

My walk to see the Red Oak begins among trails of snow fleas, delicate dotted lines spreading across soft March snow, remnants of a winter newly replaced by spring. Around me, high-pitched birdsong, a palette of gray and brown branches stippled with green buds, the distant metallic hum of some unidentified machine, and the melodious running of water over rocks in a brook invigorated by melting ice. Above, chalk-white clouds, favored by watercolorists and dreamers, wash over a broad blue-green sky. Ahead, a muddy abandoned lane, now part of Smith College's Ada and Archibald MacLeish Field Station, wide enough to move a herd of sheep from where I stand in Whately, Massachusetts, five or so miles to the town of Conway.

Perched at the top of Poplar Hill Road, the 250-acre field station is a sanctuary for intellectual inquiry, ecological scrutiny, and artistic creation. It was the province of Nipmuc, Pocumtuc, and Wabanaki communities until European settlers claimed the wide, flat ridge for their own, clearing thick forests for farms, cattle, and sheep. When the settlers' grandchildren moved west, white pine, red maple, beech, and hemlock trees took root in the lacunae where crops had grown and animals grazed. In recent decades, black birch and red oak have crowded in, too.

Wispy chilled breezes brush my cheeks. A few minutes more and I turn right, following my guide, artist Gina Siepel, off the road to another trail and toward the tree.[1]

The Red Oak is the subject of Gina's inquiry *To Understand a Tree*. A woodworker for their entire adult life, Gina's practice probes place, history, queer identity, and ecology. After a career of cutting, carving, sculpting, constructing and, at times, commodifying wood, Gina began reflecting on their "role in the ecosystem as a human, as a white descendent of European immigrants in North America, as a queer-identified artist, and as a maker of wooden objects." In response, she has spent the past three years studying the Red Oak—analyzing, experiencing, contemplating, meditating on and with, and documenting it and its local community.

The project, Gina writes, is "a small-scale way of exploring big questions about the place of humans in the ecological community, the scale and speed at which we consume living materials from nature, and how we include or exclude different forms of life in a definition of community."

Gina tries to visit the tree at least once a week. There is much to do: shooting video, writing, meditating, birding, simply being in the forest with no agenda at all. When the moment is right, she climbs eighteen feet up the rungs of a tree stand in a nearby white pine and peers into the airspace of the tree's canopy. All this effort is guided by Gina's reading of the work of Robin Wall Kimmerer—botanist, professor of ecology, and enrolled member of the Citizen Potawatomi Nation—especially Kimmerer's descriptions of Indigenous cosmologies and the ethic of the Honorable Harvest. The principles of the Honorable Harvest, Kimmerer teaches, hold humans accountable for the taking of nonhuman life. "Ask permission of the ones whose lives you seek," she writes. Then, "abide by the answer."[2]

"Minimize harm, be grateful, reciprocate the gift," Gina reflects. "Plants are kin, and we use them. *To Understand a Tree* frames attention around this relationship. How do I do that? How do we all do that in a way that is just? Ethical?"

These questions prompted Gina to look beyond natural history observations. Inspired by the Fluxus practice of "performance scores"—written instructions or prompts often implemented by audience members—Gina composed *Performance Score: Tree Breathing*. The actions listed in *Tree Breathing* activate body, brain, breath, and soul, all in connection with the tree, opening space where personal experience and art making mingled.

Performance Score: Tree Breathing

> *Part 1: at the tree*
> *set a timer for five minutes*
> *stand in front of the tree with both hands on it*
> *breathe deeply and slowly, feeling your breath fill your whole body*
> *as you inhale (draw air in through the tree)*
> *then feel your breath filling the volume of the tree as you exhale*
> *inhale (receive) and exhale (give)*
> *stop when timer ends*
> *(consider reciprocity)*

Granted permission from Smith College, Gina planned to fell the Red Oak and shape it into greenwood chairs to be exhibited along with videos revealing scenes from its life. However, three years into the project, after countless days of communing with the tree in its airspace, the form of the connection between Gina and the tree shifted, the possibilities evolved. As harvest time approached, she began grieving the tree's last set of photosynthesizing leaves and new crop of nourishing acorns. She imagined its trunk crashing to the ground and thought, *That is a bad idea.* "The tree became a teacher," she said to me. "Not anthropomorphized, not a friendly elder. A relationship that occasions learning for me."

The tree endures.

Performance Score: Tree Breathing

> *stand in front of the tree with both hands on it*
> *breathe deeply and slowly, feeling your breath fill your whole body*
> *as you inhale (draw air in through the tree)*
> *then feel your breath filling the volume of the tree as you exhale*
> *inhale (receive) and exhale (give)*

At a slim break in the woods, where lichen-covered granite peeks through snow, Gina turns left and walks off the trail. A minute later we are standing a few feet from the Red Oak, marveling at its straight, twenty-four-inch-wide trunk wrapped in gray-brown bark marked by a wabi-sabi pattern of ridges and flats resembling ski tracks and adorned with ribbons of verdant moss. To see its bare crown, eighty-five feet above us, we crane our necks. The forest's canopy is an open-work filigree of naked branches—red and striped maple, white pine, and hemlock, interspersed with yellow, gray, black, and white birch—framing bits of cerulean blue sky.

At five foot four, Gina is dwarfed by the Red Oak. Short brown hair frames their cream-colored skin and rosy cheeks, while round glasses, matching their hair, emphasize soulful hazel eyes reflecting the depth and breadth of their intellect. Ask Gina anything about the project and she will likely answer with an eloquent explanation that you'll wish you recorded to savor the pleasure of their thought and provocation.

I remember one warm afternoon, while Gina and I sipped tea on my back porch, she said: "The tree's wood is the sequestered breath of everything around it. Now and long ago. Indigenous people, colonists, farmers, students, animals and all that extra carbon humans have put into the air." This understanding, brimming with the logic of reciprocity guiding Gina's inquiry, was the yield of hours of reflective writing, observation, and tree breathing.

"Like we're made of stardust," my husband said when I read him the words "the tree's wood is the sequestered breath of everything around it," as he stirred chopped garlic and a chunk of butter into a pot of tomato sauce. "The big bang in our bones."

"Yes," I responded, distracted from his insight by the breathtaking corollary to Gina's observation. Humans are the sequestered breath of botanical life.

Performance Score: Tree Breathing

> *Part 2: at home*
> *read about photosynthesis, respiration, and carbon sequestration*
> *Increase your understanding of these processes*
> *consider the carbon cycle and its role in climate crisis*

In elementary school, I learned about the magic of photosynthesis, described by the biologist Janine Benyus as a "silent powerhouse."[3] Photons of light mixed with water and carbon dioxide and mutely mutated into sugar and starches for plants and the oxygen humans depend on. I have long focused on this gift, oxygen, but until now I did not consider how oxygen-saturated blood becomes embodied in muscle, brain, and bone—how we live, as Robin Wall Kimmerer says, "vicariously through the photosynthesis of others."[4]

What do humans offer in return? Aerial carbon, exhaled by eight billion people around the globe. Carbon, some of it respired by humans and transformed by photosynthesis into sustenance, determines a tree's magnitude and majesty. Soil secures a tree to the ground; air helps it stretch toward the sky.

After a few rounds of breathing a zesty mix of winter's decay and spring's birth, the stress of perpetual class preparation and grading subsides. My rigid shoulders soften, and I feel uncharacteristically calm. Perhaps it is getting away from campus or being in Gina's good company. Both explanations are likely, although I have been reading about a third possibility, the healing powers of trees. In 1982, the Forest Agency of Japan promoted shinrin-yoku—forest bathing—to improve the health of too overly stressed employees. Practitioners around the world now proffer the same advice. Some doctors even prescribe a trek in the woods for its therapeutic qualities.

In its simplest form, forest bathing involves walking through the woods, unplugged, engaging all five senses and breathing deeply. The practice has been enjoyed and studied by many. A recent review of mental health research in the *International Journal of Environmental Research and Public Health* attests to its power to reduce symptoms of anxiety and depression. Others extol the virtues of phytoncides—antimicrobial and insecticidal natural volatile organic compounds—released into the air by a plethora of plants. They are everywhere in the forest, and trees like the Red Oak are full of them. Evolved to inhibit or prevent the growth of attacking organisms, phytoncides service plants, but humans happily indulge in them too. They're in my husband's savory sauteed garlic, the sweet perfume of lavender and rose, the woodsy scent of cedar-lined drawers.

Some phytoncides spur the activity of natural killer cells that attack tumors and viruses. Others may lower blood pressure, pulse rates, and the hormones that wreak havoc on stressed bodies. Terpenes, found in a wide range of plants, including needle-covered conifers such as pine, may temper inflammation. Moreover, when many conifers feel hot and dry, they release terpene molecules like alpha-pinene into the air to seed the clouds, block the sun, and sometimes produce rain. Conifers can fine-tune the weather! My human-centric, Western science–trained, psychologist's mind—gobsmacked.

"Awe," says the psychologist Dacher Keltner, "is the feeling of being in the presence of something vast that transcends your understanding of the world."[5] Vastness could be physical, like the ocean viewed from the shore or an idea that boggles the mind— black holes, string theory, conifers making their own weather. A painting, song, or sculpture can feel boundless too. Whatever the spark, awe comes with questions as it challenges our perceptions of reality. Uncertainty is uncomfortable. Curiosity feels better. At its best, awe arouses fresh ways to comprehend the world and our place in it, upending coveted beliefs, making way for new ones, and deepening our sense of connection to others, human or not.

Performance Score: Tree Breathing

> *Part 3: back at the tree*
> *repeat part 1 with increased knowledge and consideration of*
> *material reciprocity*

My eyes trace the Red Oak's trunk up and through the crisscrossed canopy of featherweight branches into the sky. I feel so small. I am so small. What a relief. My day often starts with a cup of hot mint tea and a survey of the wicked problems facing my community and beyond—inequality, racism, hunger, poverty, profligate plastics, the climate crisis—and continues with a barrage of self-doubt. How can I ever make a dent? Next to the tree, I am just a little human who could do my little bit. That's all. If wood is sequestered breath, my breath adds a bit of girth to the tree, which takes carbon from the air and makes oxygen. That is something.

I exhale, long and deep, squeezing my diaphragm to push the breath from my belly into the airspace.

A week after my visit to the Red Oak, a snowstorm blows into western Massachusetts. Big flakes, weighty with water, plummet in the dark outside my window. Hallelujah. Record warmth sadly turned winter into an infrequent visitor this year in New England. I watch two pines sway, their snow-covered limbs bobbing, each tree a story taller than my neighbor's two-story home. I wonder whether these pines ever change their microclimate. Do they seed the clouds with alpha-pinene? I have lived with them for over three decades and have never noticed or cared. Now I want to smell them. I want to learn the rhythm of their terpenes.

This is what art like Gina's can do—make me sniff at trees and reconsider the composition of my airspace. A place of form and possibility, it swells with the sighs of those I cherish and have lost along with the exhalations and exhortations of strangers. It pulses with invisible molecules, their presence not quite beyond my comprehension. All a part of me, a part of us. Imagine if more people understood this. Imagine.

notes

1. As of this writing, Gina is an Arts Afield Artist in Residence at the MacLeish Field Station. The Arts Afield program, an initiative of Smith College's Center for the Environment, Ecological Design, and Sustainability (CEEDS), encourages work in, and collaboration across, the arts, humanities, and sciences at the field station.
2. Robin Wall Kimmerer, "The 'Honorable Harvest': Lesson from an Indigenous Tradition of Giving Thanks," *Yes Magazine*, November 26, 2015, https://www.yesmagazine.org/issue/good-health/2015/11/26/the-honorable-harvest-lessons-from-an-indigenous-tradition-of-giving-thanks#:~:text=The%20Honorable%20Harvest%20is%20a,between%20humans%20and%20the%20land.
3. Janine Benyus, "Biomimicry, an Operating Manual for Earthlings," *On Being*, March 23, 2023, https://onbeing.org/programs/janine-benyus-biomimicry-an-operating-manual-for-earthlings/.

4. Robin Wall Kimmerer, *Braiding Sweetgrass: Indigenous Wisdom, Scientific Knowledge and the Teachings of Plants* (Minneapolis: Milkweed Editions, 2012), 177.
5. Dacher Keltner, "Why Do We Feel Awe?," *Greater Good Magazine*, May 10, 2016. https://greatergood.berkeley.edu/article/item/why_do_we_feel_awe.

A Small Needful Fact

Ross Gay

Is that Eric Garner worked
for some time for the Parks and Rec.
Horticultural Department, which means,
perhaps, that with his very large hands,
perhaps, in all likelihood,
he put gently into the earth
some plants which, most likely,
some of them, in all likelihood,
continue to grow, continue
to do what such plants do, like house
and feed small and necessary creatures,
like being pleasant to touch and smell,
like converting sunlight
into food, like making it easier
for us to breathe.

Aura/Error

Roy Scranton

*The landscape of your word is the world's landscape. But its
frontier is open.*
    —Édouard Glissant, *Poetics of Relation*

*A note on sources: these fragments we have sheared amidst our ruin. The
intellectual's sock puppet choir, participating in a conversation across
the ages, is deployed* de rigueur, *though without attributive marking
and often inaccurately. The air flows through us in error, we err i' this
polluted air. In this, we follow our performance's urtexts: William
Shakespeare's* Hamlet *(ca. 1601) and East German poet and dramatist
Heiner Müller's revisionist* Hamletmachine *(1977).*

I. ELSINORE

Battlements. Fog.

White caltrops scatter blue channels. Wir waren Malcolm X, stand-
ing on the coast shouting at surf, while Lego Tivoli spins the kids.
The carbon party goes on: the masks we wear hide nothing, dis-
guise nothing.

Sea of fire, earth and air, this extravagant erring spirit hives
Schiphol luggage piles, Kentucky floods, jihadmall, and the team
menu is black joy, girl boss, and white ramen. The shock wave set
off in the European consciousness by the earthquake in Lisbon
in the eighteenth century has spread far and wide.... The woes
of the landscape have invaded speech, rekindling the woes of the

humanities. Can we bear ad infinitum this rambling on of knowledge? Can we get our minds off it? Here the imaginary of totality saves us… from reverted errantry.[1]

Good news! The villa's got a courtyard DJ, and you can join a Viking raid at the National Museum. The ghost on the ramparts we call COP 15, the Spirit of Christiania.

"Invisible gasses that surround the earth," from Old French *air*, "atmosphere, breeze, weather" (12c.), from Latin *aer*, "air, lower atmosphere, sky," from Greek *aēr* (genitive *aeros*), "mist, haze, clouds," later "atmosphere" (perhaps related to *aenai*, "to blow, breathe"), which is of unknown origin.[2]

"Go astray, lose one's way; make a mistake; transgress," from Latin *errare*, "wander, go astray," figuratively "be in error," from Indo-European root **ers-* (1), "be in motion, wander around" (source also of Sanskrit *arsati*, "flows"; Old English *ierre*, "angry; straying"; Old Frisian *ire*, "angry"; Old High German *irri*, "angry," *irron*, "astray"; Gothic *airziþa*, "error; deception"; the Germanic words reflecting the notion of anger as a "straying" from normal composure).[3]

Winter canceled. Spring canceled. This mourning canceled.

HAMLET
We do it wrong, being so majestical,
To offer it the show of violence;
For it is, as the air, invulnerable,
And our vain blows malicious mockery.

Enter HoratioPolonius. The TikTok Waffle House Valkyrie, her wings buffeted by a storm not so much political as theological, is propelled irresistibly backward into the future. Almost everywhere the law of blood, the law of the talion, and the duty to one's race—the two supplements of atavistic nationalism—are resurfacing…. Nearly everywhere the political order is reconstituting itself as a form of organization for death.[4]

HORATIOPOLONIUS

The air bites shrewdly; it is very cold.

It is a nipping and an eager air.

HAMLET

But soft! Methinks I scent the morning: death masks at the party...

Denmark's a prison.

HORATIOPOLONIUS

Not so, milord. Will you walk out o' the air?

HAMLET

Indeed, out o' the air—but as thought leader or gut flora, aura or orrery?

HORATIOPOLONIUS

In the field of human endeavor, milord, language is a stabilizing mechanism, not a producing mechanism.[5] "News" is mostly a tool of forgetting, a way of crowding out yesterday's headlines from the audience's consciousness. The result is the narrative equivalent of a Stockhausen score: a chain of items subject to now syntagmatic order, with no determination of later information by the preceding one, and hence a complete randomness of succession.[6]

HAMLET

Be thou a spirit of health or goblin damn'd,

bring with thee airs from heaven or blasts from hell,

be thy intents wicked or charitable,

thou comest in such a questionable shape

that I will speak to thee.

II. GAIA

Air/earth. Ofelia. Her heart is fire.

OFELIA [CHORUS/HAMLET]
This is all wrong. I shouldn't be up here. I should be back in school
on the other side of the river. Yet you all come to us young people
for hope. How dare you!

You have stolen my dreams and my childhood with your empty
words. And yet I'm one of the lucky ones. People are suffering.
People are dying. Entire ecosystems are collapsing. We are in the
beginning of a mass extinction, and all you can talk about is money
and fairy tales of eternal economic growth. How dare you!

How dare you say you hear us and that you understand the urgency.
Because if you really understood the situation and still kept failing
to act, you would be evil.

The young people are starting to understand your betrayal. The
eyes of all future generations are upon you. And I say: We will
never forgive you.[7]

I go into the streets dressed in blood.

III. HARVARD SQUARE

*Whispers and rumors. From an upright coffin inscribed HUMANITAS
steps Claudius and, painted like a wine aunt, HamletOfelia. Striptease
by HamletOfelia.*

HAMLETOFELIA
Here stalks the hope that ghosts me. *Laughs.*

*HamletOfelia spreads her legs. The angel, his face on the back of his
head: HoratioPolonius. He dances with HamletOfelia.*

HORATIOPOLONIUS
Chinblood, sitzfleisch, cranium gorged with google—oh, sweet
Prince... these seminal and, arguably, permanent constituents of
the legitimation of priestly authority have one feature in common.

They all proclaim, and explain, the separation of the priesthood from the laity.[8]

HAMLETOFELIA
My ruin-kween begged me turn up the AC, watching office lights smolder in the blue dawn from the eighth floor of our conference hotel. People rely on a limited number of heuristic principles which reduce the complex tasks of assessing probabilities and predicting values to simpler judgmental operations, and while these heuristics and related cognitive biases are necessary and useful, they often lead to severe and systematic errors.[9]

VOICES/CHORUS [*from the coffin* HUMANITAS]
What you killed you should also love.

HAMLETOFELIA
In the sixteenth century, the Florentine thinker Giordano Bruno posited an infinite universe populated with many worlds; he was later burned at the stake.... You may have ordered the chicken, but somewhere you ordered tofu, or grilled stegosaurus; a world exists in which Hillary, and the Mets, and your kid's softball team won.... Children are bringers of change and agents of chaos; they can seem to come from another dimension.[10]

HORATIOPOLONIUS
Visible nature is all plasticity and indifference, milord, a moral multiverse, as one might call it, and not a moral universe. To such a harlot we owe no allegiance; with her as a whole we can establish no moral communion; and we are free in our dealings with her several parts to obey or destroy.[11]

GertrudeClaudius enters from the laundromat above, baring breast cancer like a sun. Thousands of silver balloons stream into the sky.

GERTRUDECLAUDIUS
Invisible gases that surround the earth—

HORATIOPOLONIUS [*overlapping*]
—go astray, lose one's way.

HAMLETOFELIA
Truly, and I hold ambition of so airy and light a
quality that it is but a shadow's shadow.

GERTRUDECLAUDIUS
The obvious nightmare is that the future possibility of geoengi-
neering slows efforts to stop emissions but that the technology
turns out to be infeasible.... But there's another nightmare: It is
that after bringing emissions to zero, we realize in hindsight that
early use of geoengineering could have saved millions of lives and
helped preserve some of the natural world.... There are no easy
answers. Both errors are possible.[12]

HAMLETOFELIA
I have of late—wherefore I know not—lost all my mirth... and
indeed it goes so heavily with my disposition that this goodly frame,
the earth, seems to me a sterile promontory, this most excellent
canopy, the air, look you, this brave o'erhanging firmament, this
majestical roof fretted with golden fire, why, it appears no other
thing to me than a foul and pestilent congregation of vapors.

HORATIOPOLONIUS
Do not saw the air too much with your hand, thus, but use
all gently—

GERTRUDECLAUDIUS [*overlapping*]
—a civilization ranked Class A could re-create the cosmic condi-
tions that gave rise to its existence, namely produce a baby uni-
verse in a laboratory.[13]

HAMLETOFELIA [*swooning*]
I eat the air, promise-crammed: you cannot feed capons so.

HORATIOPOLONIUS [*catching HamletOfelia*]
Alas, how is't with you,
That you do bend your eye on vacancy
And with the incorporal air do hold discourse?
Sense, sure, you have,
Else could you not have motion; but sure, that sense
Is apoplex'd; for madness would not err,
Nor sense to ecstasy was ne'er so thrall'd
But it reserved some quantity of choice,
To serve in such a difference.

HAMLETOFELIA
Difference! The multiverse is made from trees. We never 'scape our
demons, Horatio, but learn to live inside them.

IV. BAKHMUT

Nuclear bomb. Smoking ruin of a city. Ofelia mounts an American tank
in a camo tankini, waving the Ukrainian flag. Hamlet faces Fortinbras.

FORTINBRAS
The oven smokes in peaceless October.

HAMLET
I take it Mom is dead?

FORTINBRAS
You may regret that. She brought you face-to-face with a Master of
the Mystic Arts.[14]

HAMLET
Obsessed, bewildered
By the shipwreck
Of the singular...[15]

FORTINBRAS
Denmark?

HAMLET [*activates the Reality Stone, showing a holographic image of Copenhagen*]
It was. And it was beautiful. Denmark was like most places: too many mouths, not enough to go around. And when we faced extinction, I offered a solution.

FORTINBRAS
Genocide?

HAMLET
At random. Dispassionate; fair to rich and poor alike. They called me a madman... and what I predicted came to pass. The imbalances man has produced in the natural world are caused by the imbalances he has produced in the social world.[16]

FORTINBRAS
Congratulations. You're a prophet.

HAMLET
That fabric of times that approach one another, fork, are snipped off, or are simply unknown for centuries, contains all possibilities. In most of those times, we do not exist; in some, you exist but I do not; in others, I do and you do not; in others still, we both do. In this one, which the favoring hand of chance has dealt me, you have come to my home; in another, when you come through my garden you find me dead; in another, I say these same words, but I am an error, a ghost.[17]

V. LEADERS' QUEST BREAKOUT SESSION

Hamlet and HoratioPolonius stand on the beach drinking Tuborg beer, surrounded by heaps of batteries, phones, keyboards, and smoking fires. Ghanaian children pick through the trash. Giant head of Claudius stares blind. Ofelia smiles with rue, sitting cross-legged on her beach blanket, assembling an AR-15. The sky is full of screens.

HORATIOPOLONIUS
When you first come here, it seems like a place where everything ends... The soil is dead—it's black. The river is not a river anymore. It's not water.... But that apocalyptic landscape conceals a community of entrepreneurs, determined to salvage what they can.... They see themselves as businessmen. For them it's a place of hope.[18]

OFELIA
I told them I didn't have anything to say about climate change anymore, other than I was not doing well, that I was miserable.... I told them I felt like all I did every day was try to act normal while watching the world end, watching the lake recede from the shore, and the river film over, under the sun, an enormous and steady weight. Writing is stupid. I just want to be alive. I want all of us to just be alive.[19]

HORATIOPOLONIUS
When we connect the dots, milord, we realize that climate action is not only not inconsistent with who we are; it actually enables us to be a more genuine expression of ourselves. We can become even more invested in a free economy when we realize how skewed it is in favor of polluting industries. We can believe more in personal liberties when we realize how other decisions are affecting us. And if we're a Christian who takes the Bible seriously, I think we'd be at the front of the climate movement.[20]

HAMLET
Of that I shall have also cause to speak,
And from his mouth....
But let this same be presently perform'd,
Even while men's minds are wild; lest more mischance
On plots and errors happen.

HoratioPolonius and Ofelia stare at Hamlet. Hamlet takes off his mask and costume.

HAMLET ACTOR
Enough. I'm done.

OFELIA
The idea of a demonic nature against which the self is pitted makes sense. But what about an indifferent nature which nevertheless contains in its midst that to which its own being does not make a difference?[21]

HAMLET ACTOR [*holds up his phone, taking a selfie*]
I don't want to eat. I don't want to drink. I don't want to die. I don't want to breathe. I don't want to kill. I don't want to live. I don't want to write. I don't want to think. I don't want the alienation of nine pm. I don't want to check my students' papers for the telltale signs of algorithmic intelligence. I don't want to defend the humanities. I don't want to blame STEM for the growth in business majors. I don't want to confuse air with error, or confuse tears in the rain with tears in the rain. I don't want to pretend I'm not getting old. I don't want to pretend getting old makes me wise, or foolish. I don't want to be a man. I don't want to be a woman. I don't want to be forced to say things I don't believe are true. I don't want government. I don't want anarchy. I don't want to check my feed. I don't want to complain about my timeline. I don't want to wallow in this bad air. I don't want to err. I don't want to see the auras of the damned huddled against the Walgreens in the rain. I don't want to walk down the street talking to myself. Now I walk down the street talking to myself. I see the auras of the damned huddled against the Walgreens in the rain. I err, wallowing in this bad air. I complain about my timeline every time I check my feed. I want anarchy. I want effective government. I say things I don't believe. I want to be a woman. I want to be a man. Sometimes I pretend I'm not getting old. Sometimes I think getting old is making me foolish. I confuse

tears in the rain with tears in the rain. I confuse error with air. I keep defending the humanities, and in the alienation of nine pm, I check my students' papers for the telltale signs of algorithmic intelligence. I think. I write. I kill. I live. I die. I breathe. I drink. I breathe. I eat. I breathe.

Hamlet Actor smashes his phone.

The audience on all the screens. Blood from the batteries. Three naked transwomen enter speaking at the same time, each in their own language.

OFELIA
There's scientific consensus that the lives of children are going to be very difficult. And it does lead young people to have a legitimate question: Is it OK to still have children?[22]

HAMLET ACTOR
What would you have me do?

OFELIA [*putting down her AR-15*]
For the better part of their history, Western intellectuals drew the blueprints of a better, civilized, or rational society by extrapolating their collective experience in general, and the counterfactual assumptions of their mode of life in particular. A "good society," all specific differences between numerous blueprints notwithstanding, invariably possessed one feature: it was a society well geared to the performance of the intellectual role and the flourishing of the intellectual mode of life. Each choice... was argued and legitimized in terms of the hope that the selected class would desire, and be able, to create or sustain a society comfortable for intellectual pursuits; a society which admits in practice the centrality of specifically intellectual domains (like culture and education) and the crucial role of ideas in the reproduction of communal life. No historical agent seems today to answer this description. There is no historical focus for a hope that the world might be made safe

and comfortable for intellectual work.... There is no would-be en-
lightened despot seeking the counsel of philosophers. There are
only philosophers desperately trying to create communities, and
sustain them with the power of their arguments alone.[23]

CLAUDIUS [*intoning*]
Essentially, when I was a kid I was wondering, what's the meaning
of life? Like, why are we here? What is it all about? And I came
to the conclusion that what really matters is trying to understand
the right questions to ask. And the more that we can increase the
scope and scale of human consciousness, the better we're able to
ask these questions. And I think that being a multiplanetary spe-
cies and being out there among the stars is important for the long-
term survival of humanity.... Consider two futures. One where we
are forever confined to Earth until eventually something terrible
happens. Or another future where we are out there on many plan-
ets, maybe even going beyond the solar system. I think that space
invasion is incredibly exciting and inspiring. And there need to be
reasons to get up in the morning. Life can't just be about solving
problems. Otherwise, what's the point?[24]

HAMLET ACTOR
Capitalism begins with cutting down trees.

OFELIA/GHANAIAN CHILDREN/TRANSWOMEN
Pregnant with tomorrow's swarm, I am sterile, I am fertile with
death, I am the necrophagous word hoard, I produce baby univers-
es in the laboratory, I am elective surgery, I am the open frontier, I
am that I am that I am that I am that I am that I am that I am.

HAMLET ACTOR [*kneels, hands Ofelia a gas can and a lighter*]
Your sacrifice is greater than your power.

Snow. Ice Age. Screens go dark.

FIERCELY AWAITING / YESTERDAY. *Rising seas. Exit all except*

Ofelia, who douses herself in gasoline, and the Ghanaian children, who stare unblinking at the audience.

OFELIA

What I remember most distinctly is what happened to the trees: they vanished before my eyes, melting away like snow. Ancient pinewoods disappeared along with dense spruce forests; bushes of nurseries replaced them—when, that is, they were replaced at all. Every birch thicker than a leg disappeared. Aspen groves were methodically driven to extinction.... The number of trees decreased at an inconceivable pace. I estimated that around the villages of Tavastian at the beginning of the 1980s perhaps one-third of trees were still standing that had been there in the late 1940s: a loss of about two-thirds in just thirty years. Elsewhere—particularly in the far north—the loss was even greater.[25]

Ofelia lights the match.

notes

1. Édouard Glissant, *Poetics of Relation*, trans. Betsy Wing (Ann Arbor: University of Michigan, 1997), 196.
2. *OED Online*, s.v. "air, n.1," https://www-oed-com.proxy.library.nd.edu/view/Entry/4366?rskey=q3523p&result=1&isAdvanced=false.
3. *OED Online*, s.v. "err, v.," https://www-oed-com.proxy.library.nd.edu/view/Entry/64094?rskey=EFCOss&result=2&isAdvanced=false.
4. Achille Mbembe, *Necropolitics*, trans. Steven Corcoran (Durham, NC: Duke University Press, 2019), 6–7.
5. Samuel Delany, *Times Square Red, Times Square Blue* (New York: New York University Press, 2001), 162–63.
6. Zygmunt Bauman, *Legislators and Interpreters: On Modernity, Post-modernity, and Intellectuals* (Ithaca, NY: Cornell University Press, 1987), 167.
7. Greta Thunberg, "Speech to UN," *NPR*, September 23, 2019, https://www.npr.org/2019/09/23/763452863/transcript-greta-thunbergs-speech-at-the-u-n-climate-action-summit.
8. Bauman, *Legislators and Interpreters*, 13.
9. Amos Tversky and Daniel Kahneman, "Judgment under Uncertainty: Heuristics and Biases," *Science* 185, no. 4157 (1974): 1124, https://doi.org/10.1126/science.185.4157.1124.
10. Steph Burt, "The Never-Ending Story," *New Yorker*, November 7, 2022, https://www.newyorker.com/magazine/2022/11/07/is-the-multiverse-where-originality-goes-to-die.
11. William James, "Is Life Worth Living?," *International Journal of Ethics* 6, no. 1 (1895): 1–24, 10. http://www.jstor.org/stable/2375619.

12. David Keith, "The World Needs to Explore Solar Geoengineering as a Tool to Fight Climate Change," *Boston Globe*, October 19, 2020, https://www.bostonglobe.com/2020/10/19/opinion/world-needs-explore-solar-geoengineering-tool-fight-climate-change/.

13. Avi Loeb, "Was Our Universe Created in a Laboratory?," *Scientific American*, October 15, 2021, https://www.scientificamerican.com/article/was-our-universe-created-in-a-laboratory/.

14. *Avengers: Endgame,* directed by Anthony Russo and Joe Russo, written by Christopher Markus and Stephen McFeely (2019), disneyplus.com.

15. George Oppen, "On Being Numerous," *New Collected Poems*, ed. Michael Davidson (New York: New Directions, 2008), 166.

16. Murray Bookchin, "Ecology and Revolutionary Thought," *Post-scarcity Anarchism* (Oakland, CA: AK Press, 2018), 23.

17. Jorge Luis Borges, "The Garden of Forking Paths," in *Collected Fictions*, trans. Andrew Hurley (Penguin, 1998), 127.

18. "The World's Largest E-Waste Dump Is Also Home to a Vibrant Community," *CBC Radio*, November 2, 2018, https://www.cbc.ca/radio/spark/412-1.4887497/the-world-s-largest-e-waste-dump-is-also-home-to-a-vibrant-community-1.4887509#.

19. Sarah Miller, "All the Right Words on Climate Have Already Been Said," *Real Sarah Miller*, June 28, 2021, https://therealsarahmiller.substack.com/p/all-the-right-words-on-climate-have.

20. Katherine Hayhoe, "How Do You Keep Hope amid the Climate Crisis?" (interview by Joe McCarthy), *Global Citizen*, October 13, 2021, https://www.globalcitizen.org/en/content/katharine-hayhoe-book-interview-saving-us/.

21. Hans Jonas, *The Phenomenon of Life* (Evanston, IL: Northwestern University Press, 2001), 233.

22. Alexandria Ocasio-Cortez, quoted in Nicole Goodkind, "Alexandria Ocasio-Cortez Asks: Is It Still OK to Have Kids in Face of Climate Change?," *Newsweek*, February 25, 2019, https://www.newsweek.com/alexandria-ocasio-cortez-aoc-climate-change-have-kids-children-1342853.

23. Bauman, *Legislators and Interpreters*, 147–48.

24. Elon Musk, "Elon Musk Speech: Future, A.I., and Mars," *English Speeches*, 2005, https://www.englishspeecheschannel.com/english-speeches/elon-musk-speech/.

25. Pentti Linkola, "The Green Lie," in *Can Life Prevail? A Revolutionary Approach to the Environmental Crisis*, trans. Eetu Rautio and Olli S. (Budapest: Arktos, 2011), 50.

Trespassing

Antonia Malchik

I t was late September, and the aspen trees were just beginning to yellow. They grew thick on the hillside, a broad grove giving way to small meadows that sloped upward, transitioning after less than a mile to heavy stands of spruce and pine. The group I was with rambled along an old logging road just south of the eastern side of Glacier National Park while a biologist among us talked about the ecotone we were walking through: a mingling of prairie and forest that stretched all down along the Rocky Mountain Front, the eastern-facing slope of the Rockies, where the mountains spill onto the prairie. A light wind blew constantly.

As we left the aspens and walked into evergreens, the wind became a whispering—*psithurism*, a sound that's like a rustle and a shush at the same time. That sound characterized almost my entire Montana childhood, but I never consciously noticed it until a few years ago, shortly after moving back to my hometown. One day, a few months into my return, I was walking home through town and stopped to listen to the wind blowing through a stand of tall lodgepole pines bordering the path. *That sound*, I thought, remembering its company in the Rockies on many a family hike that I had dragged my feet on as a child, and later on treks as a teenager with friends. *That sound is home.*

The place along the Rocky Mountain Front I was hiking that late September day is a two-hour drive east from the valley where I grew up. In another region, it might not be considered anywhere near my home. But this is the American West: expanses are vast, yet their very vastness and sparse human population are part of the intimate familiarity that welcomes those of us who live here. Montana is often called a "small town with very long streets." The psychological network of what I think of as my homeland encompasses the Rocky Mountain Front. For a white settler like me, a fifth-generation descendant of Montana homesteaders, the question of homeland and belonging is constantly shifting. But there is one constant: wherever my feet happen to be, my heart has always longed to be right here, among the cold mountains and prairie grasses.

Hiking along the prairie-forest ecotone, every aspect of the air felt like home—the smell of pine, the sound of wind in the evergreens, the way the sun was *almost* warm enough but the air kept me chilled. That same air had wound itself eastward from the valley I live in through a pass in the Rockies and unfurled here, to race down the foothills and speed its way across the prairie and farmland to the little agricultural town of barely two hundred people where my mother is from.

Although I never lived in my mother's hometown, or even on the kind of spread-out farmland she knows so well, the air of her childhood landscape calls to me almost as insistently as that of the stream-saturated peaks I was raised in: I can smell it now, sitting at my desk on the other side of the Rockies in a mountain valley with its different kind of big sky. I love the way virga strolls across the miles of prairie and farmland like it's got all the time in the world, how I can watch it for hours, how my skin tightens slightly at the drop in temperature, and how I can still smell the ozone of rain's promise, with its dust-tang, months later in the back of my nose. I can't understand why that air also smells like home to me, why I can look at those houses surrounded by thousands of acres of wheat and feel in my gut what it is to be a child growing up with your eyes on that far horizon, nothing between you and the rainstorm but the air and wind who make constant companions. Companions who can issue either invitation or warning, for those who listen closely enough.

There is one stark difference between these places, a difference that I too often take for granted and that most people might not notice: where I live, I'm not far from access to millions of acres of designated wilderness and national forest areas and a national park, places where my feet are as free to roam as the air itself. However, when I go out to eastern Montana, my mother's home ground, everywhere I turn is blocked by fences. You can drive for hours and see little else but weather-beaten houses huddled together on the prairie, their siding bitten with winter and the fierce, scorching sun of August. These vast counties, where you can drive past more visible wheat silos than homes and only the occasional hawk or pronghorn, are squared out and fenced off with countless miles of forbidding barbed wire.

My body can't pass through these fences without permission, but the air has no such limitations. It's a freedom that has an underacknowledged impact: No Trespassing signs are ubiquitous in America (in Montana, Trespassers Will Be Shot is a threat I always take seriously), yet at the same time, air pollution trespasses into our bodies every moment of the day. When I walk around my hometown, it's impossible not to breathe in vehicle exhaust, especially on days when an inversion layer holds it close to the ground. Out where my mother's from, on those expanses that feel like they host some of the cleanest, most unadulterated air on the planet, on any given visit I might see a crop-dusting plane emptying loads of pesticide or herbicide over the fields and still smell the strange, metallic tang in the back of my nose the next morning.

Trespass can be turned back on us. With bodies and lungs and circulatory systems porous to the air, neither humans nor the rest of life have much defense against the kinds of airborne attacks that other people have unleashed upon us. And I don't use the word *attacks* lightly. Air pollution from vehicle traffic can decrease children's lung capacity by 20 percent and significantly affect cognition in their growing brains; recently, it has been found that carbon pollution from car exhaust crosses the placental barrier and affects fetal development and even ovarian egg production in women. Living near a landfill raises a person's risk of lung cancer due to the hydrogen sulfide that's released from decaying trash. Fully 95 percent of the world's human population lives with levels of air pollution considered unsafe. Air pollution is one of the leading causes of premature death worldwide.

Without clean air, humans are denied an inherent right to health and flourishing. If billionaires' dreams of colonizing Mars were ever to be realized, the first mission, the second mission, the millionth mission, the missions for generations far beyond our imaginations would be to secure water and breathable air. Air is so vital that a common right to it was recognized in legal code as far back as the Roman Empire. "The following things are by natural law common [to] all—the air, running water, the sea and consequently the sea-shore," declared the *Institutes* of Justinian in 535 CE.[1] In 1972, after decades of relentless air and water pollution, aided by political corruption paid for by the powerful men of industry known as the Copper Kings, Montana's legislature passed a new state constitution that guaranteed a "clean and healthful environment" as an inalienable right, including the right to clean air.

Air is a shared commons: it's an entity we all rely on for survival, and it moves freely across the world. The air I breathe that smells of dry pine needles and early snow was somewhere else a few hours ago, a few days ago, a few weeks ago. Maybe it was bringing some other hikers the smell of their own woods, or picking up sulfur dioxide, nitrous oxides, and soot from a coal-fired power plant, whose particulates are now seeping into my lungs, unasked for and unwanted on a cool September day. We all depend on and all share the air, and yet the ability to pollute it is treated as a private property right. Legal systems around the world make air the recipient of industrial waste; in turn, that means that all of us are, too. Air knows no international boundaries, and neither does the pollution it carries.

When I think of trespass, what first comes to mind is the Lord's Prayer, which I recited with my parents and sisters Sunday after Sunday in Episcopal and Lutheran churches, and often around the dinner table, throughout my childhood. The lines "And forgive us our trespasses, as we forgive those who trespass against us" refer not to who crosses whose property lines but to committing sins that the deity has forbidden. The word *trespass* occurs many times in the New Testament. In some translations it's replaced with *sin* or *debt*.

Trespass, in other words, is a transgression. In the case of pollution, trespass is far more invasive than simply breaking through a property line. If I sneak through my neighbor's yard to get to the public nature preserve on the other side, I might annoy him, but there's no actual harm done. If my neighbor burns a pile of tires in that same yard and I don't go near it, his waste will trespass into my family's bodies just the same, pouring itself into my children's lungs with the law's consent. The polluted air has trespassed into us, but it wasn't by choice. The first crime of trespass was against air itself. When air has been violated, it is forced to violate in turn.

I was hiking along the Rocky Mountain Front in late September 2022 with a group working to stop oil leases in what is known as the Badger-Two Medicine. It's an area bordered by Glacier National Park to the north, the Bob Marshall and Great Bear Wildernesses to the west and south, and the Blackfeet Reservation to the east. The Badger-Two Medicine is a sacred place to the Blackfeet Nation. Under laws written and enforced by the federal government, it's legally part of the US National Forest Service, but it was carved off of Blackfeet land in 1895, along with the eastern part of Glacier Park, in yet another land seizure accomplished with a deceptive treaty signed under duress, one in a long history of betrayals.

I hadn't been to the Badger-Two Med before, although I'd been following the oil lease situation—which has been ongoing for nearly forty years—since before moving back to Montana. This was the first time I'd managed to visit it, on a hike sponsored by the Glacier-Two Medicine Alliance, which was founded in the 1980s to fight oil leases granted in the area by the Reagan administration. Most of the leases have been successfully canceled over the years, but when I first hiked there in late September 2022, one oil company had just won a court appeal to keep its lease.

Emerging from the aspen groves and into pines and spruce, my group walked a path that ran parallel to a buried natural gas pipeline; the organizers pointed out where a road to the remaining proposed site of the oil well would be built if the lease were upheld. A few miles further in, we would see a hillside already scarred by preparatory clearing.[2]

It's hard to imagine a place that feels more like the white European settler's idea of pristine wilderness. Pristine wilderness and its ideals of unchanging purity have never really existed, of course, but perhaps places like this offer something better: I felt whole on that hillside. The air's movement and scent felt like a welcome. And even though I know that there is no clean air, really, anywhere in the world—everything from dioxins to Chernobyl radiation has been found in polar ice, carried by the air and dropped even on places where few humans have ever stepped—I felt an extra surge of resentment at the thought of the trespass that would come not just from the physical invasion of an oil well but from the particulate matter, carbon monoxide, nitrous oxide, and volatile organic compounds that have been found in the air around and downwind of oil-drilling operations. At what the air would be forced to carry through no choice of its own.

As we walked to the top of a hillside where we could see out toward the plains of eastern Montana, the air shifted from a gentle breeze to a wind traveling east—stiff, but not quite the hard-blowing kind that is almost a constant presence on the wheat and cattle ranches that cover what's known as the Golden Triangle, the wheat farming region my mother grew up in.

The wind blew the smells of encroaching autumn in my face, dried grasses underfoot and fecund soil under bear-claw-scarred aspen trees. The tiny bit of late-September chill reminded me that snow would be coming soon. There is nothing that smells more alive to me than that air. It feels conscious: the warm pine in summer, the tang of ice in winter, traveling down from these mountains to kick prairie and dirt-road dust in the faces of children growing up in the same tiny town my mother had over seventy years before. The heart that has always insisted on calling this place home, even during the twenty years I lived elsewhere, tells me, quietly, that this air I love in all its moods and seasons is conscious. It has a life of its own and a right to live it unviolated.

The crime of trespass goes both ways—what happens when we require the very source of life to carry sickness instead? Is this not a violation of the gods of life, of home, and of air's own right to exist?

Acquiescence to the abuse and neglect of air is a trespass against humanity—against all of life, even against the air itself, for its own sake. Every living being has a common right to air that not only allows us to live the healthiest lives we can but also smells like pine and snowmelt, desert dust and prairie flowers, swamp grasses and moss. Air that feels like home.

notes

1. Justinian, *Institutes*, bk. 2, title 1, "Of the Different Kinds of Things," trans. J. B. Moyle (Oxford, 1911), available at https://amesfoundation.law.harvard.edu/digital/CJCiv/JInst.pdf.
2. Almost a year later, in September 2023, that lease—the last in the Badger-Two Medicine—was bought out and is finally being retired.

Inebriate of Air

Benjamin Kunkel

1.

For the longest time, air was invisible. You simply didn't see it; you just saw *through* it.

This was true, first of all, in a literal sense. Air was by nature invisible, casually and totally so, except on such occasions as when you (or, anyway, I) surprised it into view by climbing up the red slope across the road and looking down the little valley into the haze of half a dozen miles' distance or so; or else when a parent stopped the car at some roadside point of interest in the mountains and then the film of graduated dullness laid out across the receding ridgelines disclosed the physical substantiality of the otherwise bodiless atmosphere you inhabited.

Otherwise, though, you took the apparently immaculate air for granted. No thing itself, it was instead the transparent medium in which all other things existed or took place.

And even less than you (or, again, I) literally saw the air back then did you ever really think about it and see it in your mind's eye—as something perhaps delicate and at risk, or dangerous and unreliable. Even when, as an adult in my thirties, I lived in the enormous traffic-swarmed city of Buenos Aires and almost every day found myself walking down this or that narrow street behind the tailpipe of some colorful city bus, hardly ever did I pause to consider, except in maybe the most glancing way, the issue of "air quality," much less acquaint myself with the measure of the Air Quality Index (AQI). After all, the flat expanse of Buenos Aires lies at the far edge of the planar pampas, not in some basin where smog might linger; and even at those times when the fair winds after which the city is named took a few days off and you could smell the traffic, a thunderstorm would soon enough roll through and restore the air to something close to cleanliness.

2.

I'm trying to say that the former glory of the air, it seems to me now, was to be entirely overlooked—like the self-effacing God of some atheist people who depended for all they had on their god's unsuspected grace.

Of course, to say as much is partly just to record my own naivete and shelteredness. I grew up in a pair of narrow mountain valleys on the western slope of Colorado, except for a spell of a few years in what seemed to me the very big city of Glenwood Springs (population about five thousand at the time). It's true that the TV would speak of "the brown cloud" of air pollution that often settled over faraway Denver in the 1970s and 1980s—and that might have clued me in to the fact that the blessing of clean-enough air was far from universal, and that even in a rich country like the United States there were, for instance, neighborhoods and cities in which kids, especially poor kids of color, suffered from disproportionately high rates of asthma. And eventually I even learned these things. Nevertheless, the air, like I say, rarely entered my mind. Until a few years ago, I suppose I basically thought: Air quality has improved a lot in the United States since the passing of the Clean Air Act in 1970, and stricter car emissions standards later on, and among all of our problems, this is probably one that we don't need to be worrying about too much.

Today, things have changed. For me and, I suspect, for lots of us now—here in another of the smoke-filled summers of the early 2020s—air is no longer invisible. I read the articles about indoor air pollution, especially from gas stoves—and I switch to an induction range. Then I read the articles about ozone pollution where I live now, on Colorado's Front Range, which for nineteen years in a row has exceeded Environmental Protection Agency standards for safe levels of ozone—and I purchase an electric vehicle. And then I read the articles about the effects of wildfire smoke on one's health— shortened life span, reduced lung capacity, increased risk of heart attack and depression (tell me about it)—and I order an indoor air purifier, not to mention skipping my morning hikes or runs on days when the AQI is particularly high. Finally, I read the articles on global warming and see that capitalist civilization has now overshot the boundary of 350 parts per million of carbon in the atmosphere—a level at which this civilization might survive—by approximately 65 parts per million. And for this I have no private remedy.

Indeed, I'm part of the problem. In spite of my vegetarian diet and electric car, I'm a citizen of a country whose inhabitants emit more carbon per head than those of any other large, rich country, except Canada. The climate marches I've gone on have done more for my conscience than for the planet. This past summer, a book I coedited, *Who Will Build the Ark? Debates on Climate Strategy* from New Left Review, was published. Nothing could be less likely than that it will acquire as many readers as Glenwood Springs, Colorado (population now ten thousand plus), has residents. And if the book does add its little quantum of momentum to the gathering international climate movement? Governments today, including mine, are better at nothing than ignoring mass protests.

The air has more or less become my visible enemy, and I, its. I harm it all the time, and it does the same to me and my lungs.

3.

"Inebriate of air am I, / And debauchee of dew," Emily Dickinson says of herself in one poem.[1] Something similar seems to have become true of me, even as the breathing atmosphere every day darkens and decays with fatal particulates and intimations of collapse.

This is a fancy way of saying that, since moving back to Colorado in 2018, I've done a lot of hiking. It's not like I'd never laced up a pair of hiking boots before: In Argentina, I sometimes took flights to the Andes and tromped around amid the disappearing glaciers; and in New York City, I occasionally boarded the train to Cold Spring and hiked to some perch over the Hudson. But walking up and down mountains wasn't the regular part of my life that it's become over the past five years.

In the winter, my friend Samir and I try to ascend Green Mountain (8,148 feet) every week and half-succeed; we probably average every other week. And over recent years I've heaved myself up nearby Mount Sanitas (6,863 feet), Bear Peak (8,459 feet), and South Boulder Peak (8,549 feet) more times than I can count. But these are dwarf summits by regional standards. My favorite time of year arrives around the first of July, when the high country is free of snow, and you can ascend the highest peaks—12,000 feet and higher—without gloves, goggles, balaclava, crampons, and so on. Nothing but stray rags of snowfields remaining up high, you climb through pines, wildflowers, tundra, and talus. The sunlight as you go is so pure, abundant, and crashing that it awes and somewhat frightens and always gladdens you. This summer, the tallest mountains I went up were Mount of the Holy Cross (14,009 feet), Mount Sherman (14,043 feet), Missouri Mountain (14,074 feet), and Blanca Peak (14,351 feet), the last of these the tallest point in the Sangre de Cristos and the fourth-highest summit in Colorado.

A few years ago, when I dragged my partner Hermione up Mount Audubon (13,229 feet) in early September, she—someone in very good shape, who does yoga or goes running nearly every day—described this as perhaps the worst physical ordeal of her life, followed by an almost unprecedented feeling of elation. It was like doing drugs in reverse, we decided: misery succeeded by bliss rather than the other way around. And it seemed like a harmless, clever comparison at the time.

More recently, the comparison with drugs has come to seem less innocent to me. I want access to the thin air constantly, compulsively, and will hike up a tall mountain while the season allows even if it does me harm. A podiatrist has been treating me for plantar fasciitis. At my last consultation I told him that I'd given up running for now, as he'd suggested, but hadn't been able to stay off the peaks. "How big are we talking?" he asked me about my last hike. I said: "Fourteen miles, about 5,500 feet of elevation gain." He shook his head and said I wasn't going to get better until I gave my feet some rest. And so I promised—like a drunk circling a future date on the calendar—to get sober and stop hiking for a while on November 1, when the peaks will be snowed in.

At 14,000 feet, there is more than 40 percent less oxygen than you enjoy at sea level, and maybe it's the scarcity and purity of the air up there that lets me forget what the air is to me otherwise, down below. I concentrate on my breath, and my solitude. I hardly think about anything but the few steps in front of me. I'm not in a car, or at a computer, or at home with the air purifier and electric range. In spite of the shortage of oxygen, there is just enough of the stuff: it's so hard to breathe that mere breath is a joy. On good days, most of the haze in the air is water vapor, not wildfire smoke. My skull rings like a bell, and I'm grateful as a lottery winner simply to be planting one foot in front of the other across the jumbled rocks.

But you can't live up there, in the thrilling scantness of O_2. You have to come down. You have to descend back into a choking world busy, including in your activity, at its own death.

There is no note of resolution to end this on.

notes

1. "I taste a liquor never brewed (214)," in *The Poems of Emily Dickinson*, ed. Ralph W. Franklin (Cambridge, MA: Belknap Press of Harvard University Press, 1999), https://poets.org/poem/i-taste-liquor-never-brewed-214.

Ozone

Rita Dove

...Does the cosmic
space we dissolve into taste of us, then?
—Rilke, *The Second Elegy*

Everything civilized will whistle before
it rages—kettle of the asthmatic,
the aerosol can and its immaculate awl
perforating the dome of heaven.   .

We wire the sky for comfort;
we thread it through our lungs for a perfect fit.
We've arranged this calm, though it is constantly
unraveling.

Where does it go then,
atmosphere suckered up
an invisible flue?
How can we know where it goes?

A gentleman pokes blue through a button hole.

Rising, the pulse
sings:
memento mei

The sky is wired so it won't fall down.
Each house notches into its neighbor
and then the next, the whole row scaldingly white,
unmistakable as a set of barred teeth.

to pull the plug
to disappear into an empty bouquet

If only we could lose ourselves
in the wreckage of the moment! Forget
where we stand, dead center, and
look up, look up,
track a falling star...

      now you see it

        now you don't.

An Invitation to Lose One's Way: In Conversation with Báyò Akómoláfé

Báyò Akómoláfé and Daegan Miller

W hat might it mean to live an airy life? How does dis-ciplinary knowledge hem us in? How can getting lost help us to find where we are? These are some of the questions that the clinical psychologist, father, international speaker, public intellectual, and writer Báyò Akómoláfé follows in his genre-bending work. Akómoláfé is the visionary founder and elder of the Emergence Network, the author of *These Wilds beyond Our Fences: Letters to My Daughter on Humanity's Search for Home*, and the editor of *We Will Tell Our Own Story: The Lions of Africa Speak*.

In May 2020, while the world was reeling from COVID-19, Akómoláfé published a... Well, it's hard to know what to call it, exactly. Part theoretical analysis, part time-traveling science-fic-tion adventure, and part first-person meditation on family and activism, the gorgeously illustrated—let's just call it an essay—"I, Coronavirus: Mother. Monster. Activist." showcases the fugitive complexity of Akómoláfé's thinking and aesthetic practice.

In the following expansive interview, I asked Akómoláfé to extend his thinking from the monstrous virus to the element air, and his responses—layered and at times counterintuitive in their flow—highlight the possibilities of what he calls "beautiful cracks," spaces where the air gets in.

Daegan Miller: So much of your work, as I read it, and especially in the essay "I, Coronavirus," is about decentering the myth of the individualistic, unitary "master of all he surveys" capital-*M* Man, the erection of which is one of the grand projects of the European Enlightenment.[1] One of the ways you do this destabilizing work in "I, Coronavirus" is by personifying the COVID-19 virus as an abandoned, vulnerable little girl—but an otherworldly, slightly monstrous little girl. What makes monsters, monsters is that they're neither entirely their own beings nor entirely our own creations but some mix of both, like Frankenstein, who was partly the creation of Doctor Frankenstein but also his own being, with a volition and agency not entirely controlled by his maker.

Your essay asks us, echoing Bruno Latour, to love our monsters.[2] Why? What would it mean to do so?

Báyò Akómoláfé: I grew up in Nigeria, and as children, we were told monster stories. These stories always ended with a moral, a lesson to be learned. And the lesson almost always to be learned was "Stay in the moral territories that have been assigned to you. Stay away from the monsters, those figures at the edges. That is how you have a good life. That is how you raise good children and have a good future with plenty of yams in your barn. And that is how good things happen to you." If you stay within these moral territories that have painstakingly been marked out over generations, then things will be good for you.

Monsters are the edges; they are the embodiment of the places where we're not supposed to go.

And I think every act of building a settlement is an act of creating moral territory, right? Every act of settlement building is an act of codification, of moralizing and making moral imperatives. This is unavoidable and often protects us.

But there are times when doing so becomes counterproductive: the ritual doesn't *take* any longer, the terrain is constantly shifting, amenable to forces beyond our codes, our language. Then the rituals and the codes and the language and the syntax and the grammar of civilization actually become toxic in their proliferation of stuckness. There isn't mobility. There isn't imagination. There isn't thought. It's just repetition.

I sense that it is in those moments that errancy is required—*errancy*: deviation, disruption. Many indigenous cultures speak about the archetype of the trickster. You cannot conceive of emergence or movement without noticing the gift of the trickster, which is disruption.

This is where the monster comes in. The monster is not just the embodiment of the edges; the monster is also the embodiment of the crossroads. This is how the Yoruba people think of monsters: they are a crossroads. The monster is an invitation to lose one's way, a repudiation of the forms that we've adopted in order to move and to live and to be recognizable and to be legible within civilizing moralities.

And so the monstrous is a calling out of form. It's a calling out. The monstrous is a *nah, nah, nah*. You touch your seams, you touch your edges, and you find out that you're not as well put together as you think you are. That's where embracing the monstrous becomes an invitation to novelty, to what I would call sensorial mutiny, or some kind of ontological apostasy.

Daegan Miller: I want to leap off from your own language here to think about air. It's not quite monstrous, but it is sort of a trickster. There's a nice homophony in *air* and *errancy*, and I love this idea that we are not put together as well as we like to pretend that we are, that our culture isn't as monolithic as we like to pretend that it is. How do the elements, how does air, get in between these seams, these edges?

Báyò Akómoláfé: Well, I'm going to play with this a little and think about the elemental or the elements as enlistments of sensorial apparatuses.

There is a sense in which we tend to think of the elements as fixed principles, fixed things in the world. And I wonder if this fixity isn't a coproduction of our human bodies, a production of the stories that we tell, contingent upon certain ways of being in the world. Air seems so obvious, and water seems so obvious, but what if we took on different forms—for instance, if we took on a hybridized alien form? Or became cyborgs? What stories would a dust mite tell about air, or the demodex, which is a microscopic cousin of the spider that lives in our hair follicles? Would air have the same quality that it seems to have now, to you and me?

Air can be only a contingent reading at a particular moment in time.

But air in itself, the element in itself, spills beyond legibility. It's more than just a vector for molecules or particles or ideas—or for spirits or principles. This *beyond legibility*, this beyondness, this spillage—it gives us life, right? And yet the thing that gives us life is not quite a thing. It's movement. It's promiscuity. It's the material promiscuity of a world that cannot be stilled.

So I think that when we are investigating air, we're not investigating some external Aristotelian principle that is just out there for objective analysis. We're instead calling ourselves and our entanglements with the world into question. We are inviting inquiry into politics, into the ways our bodies are framed, into the ways that we are concatenated with systems, with ecologies, with environments.

Daegan Miller: One of the things that's been so interesting for me to watch as the editor of this volume is how every writer struggles with the ineffability of air. Air is a hyperobject, right? You can't point to it. You can't point to something that's not air. Air permeates most things, and the air is filled with all sorts of things that aren't air. We could say that air is everywhere and everything, but that's not all that helpful, almost like saying air is nothing at all.

As writers, we make forms. This, then, is the struggle: how to give form to something that seems like it wants to remain illegible, formless? There is, of course, no one solution to this struggle, as the essays in this volume elegantly attest, but I want to put the question to you directly: how do you reconcile the need to bring definition to the world in a way that honors the world's resistance to definition?

Báyò Akómoláfé: I think that act of noticing that the world spills beyond our definitions is already a beautiful crack in the modern, colonial epistemology. There isn't some watertight way to approach the world, to categorize it in with finality, to archive it, put it on the shelf and say "Dusted and done." And yet we must struggle to make sense of the world—I think that struggle is redemption.

Daegan Miller: Let's linger on this note of redemption for a moment. There's a scene in "I, Coronavirus" in which the main human character, Braveheart, a psychologist sent to interview the little girl, COVID-19, is transported back in time by the girl to the hold of a slave ship somewhere in the Atlantic in the Middle Passage. Braveheart leaves the hold and its "sickening smell of excrement and piss," ascends to the main deck, and has a moment of near communion with heavenly eternal bliss. But the bliss fades, he's left bereft of hope, and has an epiphany: he must return, voluntarily, to the hold, with its pestilential, fetid air. "This ship," Braveheart realizes, "riven with cracks and pain, is my mother, and I must descend into her womb to know her in a way that resolution could not offer."

It is a beautifully rendered, complex, brutal scene. I'd love to hear you think out loud a little bit about that moment of turning in to the despair, turning in to the rank, miasmatic air, the decision to voluntarily return belowdecks. "Trust in this brokenness," you write.

Báyò Akómoláfé: I remember, at the time I was writing this, I was sitting with the bones of an enslaved woman—Bakhita, they called her, named after a twentieth-century Sudanese-Italian saint—at the old wharf area, Cais do Valongo, in Rio de Janeiro, Brazil. Cais do Valongo received more black bodies during the slave trade than any other part of the world. Of course, many people didn't make it. Bakhita didn't. Upon arrival, she was dead and buried in a borderland—I should cancel out the word *bury*, because she wasn't buried at all. She was stuffed with discarded pieces of pottery, just squeezed into a place in the ground.

In our own time, a woman named Mercedes Guimarães bought herself a home. She did not know that it was built over Bakhita. She had had some renovations done on her home and discovered bones in her yard. The anthropologists came, and they named the bones Bakhita, and they said Bakhita was a twenty-year-old woman when she died.

Guimarães transformed her home into a museum dedicated to the memory of the slave trade, and it's the only one in Brazil. Her house is a UNESCO site now, and she refused to cover the graves, to cover the holes. So you can literally see the bones below; you can see Bakhita curled like a mango of some kind. I sat by Bakhita, and while sitting with Bakhita, I think I may have received an invitation to rebuild the slave ship.

Rebuilding the slave ship is not exactly a thing you tell people to do because of what it revives! But rebuilding the slave ship became some kind of research inquiry. In rebuilding the slave ship, we are touching the contours of our captivity, and we are noticing that disembarkation never happened. Disembarkation is the myth, the modern myth that the slaves got off the slave ship and the slave masters got off the slave ship.

The slave ship did not just disappear into history; it vomited out its guts, and those guts, its entrails, became the machinations and mechanisms and algorithms of our modern life. We are all part of the slave ship, and we never left. None of us has left, although we're not all in the slave ship in the same way.

I want to touch the trouble, stay with the trouble. Staying with the trouble means noticing. So why have my character go back down into the hold of the slave ship? Let's think about our modern politics. A lot of contemporary politics is framed around inclusivity, diversity, equity, representation—like coming up from the hold for air. The captains of the slave ships would bring up the preferred slaves and get them dancing on deck so that they could breathe the fresh air and get some exercise and sun. This was a way of preserving those enslaved people, a way for the captains to protect their economic investment.

That upper deck of the slave ship now lives on in the form of our politics of inclusion. Just as getting enslaved people dancing in the sun wasn't about re-forming culture or politics, so too is our politics of inclusion preserving our violent culture rather than creating something new. Our politics of inclusion is the afterlife of the upper decks. The problem with this is that you're still part of the furniture of the slave ship.

Being dragged up into the air and voluntarily going down back into the hold are two ways of responding or being in response to air. The rank, dank, horrible, steamy, noisy quality of the air in the hold is horrible, terrible. But it's also where the trickster lives. The trickster lives in those moldy places. Those moldy places are our crossroads, our invitations to wander away from the conventional path.

You know the word *nitty-gritty*? We use it to mean "getting right down to facts." Do you know where the word comes from? It means the air that comes up from the hold of the slave ship. And yet we use it today to mean getting to the very heart of things.[3]

Let's do that: let's get to the heart of things. Let's get to the facts. The air in the slave hold is horrifying. It is monstrous, but it's also the holding place for Creole futures, like the trickster. If the stories of Yoruba elders are anything to go by, the trickster was a stowaway on those slave ships. He lived in the air of the slave hold. He stayed there. He didn't go out on the upper deck. He was there, right there in the horrible air of the Middle Passage.

Daegan Miller: I want to end this conversation on the act of claiming sanctuary, which is central to your thinking. You note that, in its medieval and early modern European roots, in order to claim sanctuary, an outlaw had to hang on to the sanctuary knocker of a church, and the knockers were often fashioned to look like monsters. Sanctuary literally involved grabbing hold of the monster, and doing so allowed the outlaw—a boundary figure, like the trickster or the monster—space and time to breathe freely.

How do you think about sanctuary in our colonial, racialized, Anthropocene age?

Báyò Akómoláfé: I used to think about making sanctuary as the work of protecting excluded people, people in troubling situations, refugees, immigrants. And these are important things to do, especially now.

But I also think that I was blinded to seeing, I was blinded from noticing, that the one who needs sanctuary the most is the monster, that the subject of sanctuary isn't the excluded subject per se.

In times of shifts and dramatic movements, the world proliferates monsters. We get back to normal by killing these monsters, right? By suffocating them so that we can restore the moral edges and territories that we're used to. But in times like these, when the moral edges and territories are no longer useful, the only way for us to thrive is to make room for the monster, to allow the monster to breathe.

That image of the sanctuary knocker: maybe that's why it looked monstrous. It's almost saying, "Are you a monster? If you're a monster, you're welcome here."

This always bothered me. Why would a monster be the herald of sanctuary? And then I realized it's not about granting protection. It's about accompanying the monstrous in its emergence, in its playing out of itself.

Making room for the monster, accompanying the monster: this is my politics. And that's very practical, not esoteric or transcendent.

I remember that once my son, who is autistic, was having a moment, and I, as a psychologist, have been trained to think of these as tantrums or meltdowns. My wife has forbidden that language in our home; she prefers that we call these moments "the passing of a wild god." My son was having one of these moments when we were in a shopping mall in Chennai: we're just walking around, being good consumers, doing our capitalist business. And something happened to him, he was five or four, and he was suddenly screaming. He was in a bad place. I remember trying to resuscitate him by consoling him, pulling him up to the deck, restoring sanity, trying to drag him into the boundaries of the known, and of the common, and of the legible. I remember offering him a cookie, trying to shush him up. It was getting embarrassing. Eyes were already on us, and I'm a black guy in Chennai. I was thinking about all the other implications, and he was really not having a good time. I was also failing as a father in that moment.

I remember my wife walked up to us and said to me: "Walk away. I'll take care of this." And she got down by his side right there on the floor, lying next to him, side by side. She didn't say a word to him. She refused to give him hope. She refused to cuddle him. She just accompanied him. She wanted to stay with what was happening.

Maybe what we need to do is to accompany our monsters. Maybe the thing to do is to get beyond language or come out from the other side of language so that we can sense the world differently. Maybe we will notice wonderful things if we do.

I think accompanying needs to be the basis of the politics of our time. Making sanctuary is accompanying the monster in its errancy and noticing that we ourselves are not well put together, that we are concatenations of desire in its flow. And the politics of our time is to flow with desire to see where it might take us.

notes

1. Báyò Akómoláfé, "I, Coronavirus: Mother. Monster. Activist.," https://www. bayoakomolafe.net/post/i-coronavirus-mother-monster-activist.

2. Bruno Latour, "Love Your Monsters: Why We Must Care for Our Technologies as We Do Our Children," *Breakthrough Journal* 2 (Fall 2011): https://thebreakthrough.org/journal/issue-2/love-your-monsters.

3. There is a good deal of debate as to the etymological origin of *nitty-gritty*. The *Oxford English Dictionary* traces the birth of the word to mid-twentieth century Black American slang, with no explicit mention of slavery. However, arguments over the phrase's origins have surfaced periodically in the British press since the early 2000s. See, for instance, Anthony France, "BBC Rejects Complaint against Laura Kuenssberg for Saying 'Nitty Gritty,'" *The Standard*, January 23, 2021, https://www.standard.co.uk/news/uk/bbc-laura-kuenssberg-nitty-gritty-b900886.html.

American Atmosphere

Craig Santos Perez

June 23, 2020

I open the sliding glass door to air out
our quarantined apartment. The trade winds
are dying, and the sky is the bluest I've ever seen.
No planes overhead, fewer cars on the freeway.
They say, the ozone layer is healing, pollution
at an all-time low. Yet thousands will succumb
to COVID-19 today, another black man killed
by cops, another suspended from a tree in strange
weather, another peaceful protest dispersed
by tear gas. Plumes of dust from the Sahara desert
storm towards us. Will it trigger my daughter's
asthma? How do hospitals decide who gets the last
available ventilator? Whose lives matter enough
to breathe in this suffocating atmosphere?

To Bring Down a Sky

Gabrielle Bellot

I n the sixth century, as the stupefying mists of the Dark Ages were beginning to fill Europe, a trio of anchorite monks decided to wander the Earth until they found Heaven. The monks, who were devotees of the early Christian Saint Macarius, traversed a series of ever-more-fantastical lands until, suddenly, they came to a part of the world where the sky was so low and close to the ground that they had to stoop. When they crawled further, they found that they could no longer tell the ground apart from the sky, and, by peering as if through a celestial curtain, they could see "beyond" the empyrean, they imagined, into Heaven itself.

The improbable tale was carried across Europe by the Crusaders, but it might have been forgotten if not for a series of references, centuries later, by the avid astronomer and balloonist Camille Flammarion, who frequently alluded to the boundary-blurring adventures of those ancient anchorites in his books. The myth was memorialized perhaps most notably on the cover of Flammarion's popular publication *L'atmosphère* (1888). In jeweled colors, it depicts a man crawling through a point at which the sky seems to have curved so sharply downward, so near the ground, that the man needs to be on his knees to pass under it; in the image, the man has pushed his head straight through this low sky, like a worm poking out of an apple, staring up at wonders beyond anything native to either ground or sky. The image, which came to be known as the "Flammarion Engraving"—and its prostrate protagonist, the "Flammarion pilgrim"—resembled a medieval woodcut but was likely designed by Flammarion himself.

In some ways, the picture, despite its air of aesthetic medie-valism, was especially germane to Flammarion himself. An author, astronomer, and engraver, as well as an ardent enthusiast of hot-air ballooning, Flammarion—who, with his wild hair and deep pensive eyes, seemed the prototype of a Romantic nineteenth-century mad scientist—had long been drawn to the skies, whether by gazing through telescopes or from the confines of high-flying balloons. He had first fallen under the spell of aeronauts, or voyagers of the air, at the age of sixteen in the Jardin du Luxembourg, when he saw a balloon pass above him, the man and woman in it waving down at him from the celestial heights. "I would have given the world to be in the car of that balloon; and long afterwards I could think of nothing but a journey *into the atmosphere*," he said.[1] He often waxed poetic about the air, his romantically tinged rhapsodizing perhaps influenced by his early training as a priest, and once he began his many rides in the sky, his love of the heavens only increased.

"This marvellous world of air, so mild and yet so strong," he described in "A Sketch of Scientific Ballooning," an 1867 essay, as if he, too, had poked his head through a perspectival curtain and witnessed the heavens for the first time. It was perhaps inevitable, then, that he would turn to a picture of someone dissolving the boundaries between earth and air, the mundane and the sacred, the known world and the mysteries of the universe at large. His accompanying description of the image linked the air above to mystical speculations and alluded to the tale of the monks (here represented as one monk, although the legend describes three):

Whether the sky be clear or cloudy, it always seems to us to have the shape of an elliptic arch; far from having the form of a circular arch, it always seems flattened and depressed above our heads, and gradually to become farther removed toward the horizon. Our ancestors imagined that this blue vault was really what the eye would lead them to believe it to be; but, as Voltaire remarks, this is about as reasonable as if a silk-worm took his web for the limits of the universe.... A naïve missionary of the Middle Ages even tells us that, in one of his voyages in search of the terrestrial paradise, he reached the horizon where the earth and the heavens met, and that he discovered a certain point where they were not joined together, and where, by stooping his shoulders, he passed under the roof of the heavens.[2]

The world above us, as his silkworm analogy suggested, was far vaster and more complex than we could imagine—a comparison no doubt appealing to balloonists like Flammarion, who, after all, had gained a brand-new perspective on Earth from the heights of a flight. It is fitting that Flammarion's engraving offers only a vague glimpse into what that stooping "missionary" glimpsed once he saw beyond his own web: an enigmatic indigo space of what appear to be clouds, perhaps a bit of jagged nebulae or aurorae, giant stars, and what appears to be a colossal spherical tool, all separated by lines or layers—with a suggestion that much more lies beyond what the image shows.

It was, in other words, a beautiful, byzantine mystery, composed as much of things we might find familiar as of inscrutable arcana. Flammarion's engraving obviously does not move, yet there is an implied motion in the image, a curving downward of sky and stars for the pilgrim to crawl under. There is a sense of pure enigma in the strange blue depths he gazes up into, a dance of things that defy language, as sudden and animate as the aurorae that shimmer so cryptically above the clouds on certain nights.

I first saw the extraordinary picture many years before reading about those legendary monks, and at the time, I, too, thought it was a fetching, if esoteric, medieval rendering. Recently, though, I saw it again by chance, and that was when I learnt the story of the Macarian wanderers. I kept thinking of the idea underlying picture and tale alike: the sky is divided from the ground, but at some point, the two merge. I loved, too, the vividness of color of what we could see set against the blue of the enigmatic world beyond the "roof of heaven." The sky, and the cosmos as a whole, brimmed with the suggestion of life in his engraving.

I thought of the image again earlier this year, when my wife and I took a brief trip to Iceland. A lot had changed in our lives suddenly. I had been laid off from my job with no warning or clear explanation just a few weeks before we were set to travel; for a week, I cried and raged intermittently at the unceremonious unfairness of it all. At almost the same time, my wife learnt that her father, who lives in a Chicago suburb, had been hospitalized after a frightening incident, and we found ourselves in an emotional maelstrom, each trying to be there for the other when we weren't ourselves overwhelmed. We wondered aloud whether or not we should, or could, still go on the Icelandic trip we had been dreaming about since the end of 2022.

Eventually, we did, deciding that we, like those monks, needed to go find a bit of heaven on Earth. We left in the evening from Queens and arrived in Keflavík in the wee hours of the morning. Even before the plane landed, the empyrean was on everyone's minds, for the flight attendants assured us that we should look outside a window because, just the night before, they had seen the Northern Lights. Indeed, as I would learn, the previous night had been one of the most extraordinary nights for the Lights in a while, with shimmering clouds of green-red dancing over the skies as far away as Scotland and upstate New York. Our flight wasn't lucky, though, and when we landed, all was a vast, faintly cloudy, predawn black.

Still, even if we didn't see the aurorae from the airplane, the sky over Iceland was astonishing. Everything felt vast, present, powerful in a way I couldn't look away from. I hadn't expected just how *much* of the heavens there would be relative to the skyscraper-studded skies of New York. The Icelandic sky was vast, so vast that on some days one side was a dreamy orange-pink and the other a wildly shifting

gradient of blues; look up around a bend and you might see, sud-
denly, a fleck of rainbow hovering above the edge of a snowcapped
mountain, like a scrap of comet tail frozen in the air. A rainbow I
saw near Gullfoss, a gargantuan waterfall, seemed utterly mythic, a
Brobdingnagian colored arch just *there*, which most people glanced
at for a second or took a quick photo of and then seemed to forget
about. I stood staring at it, feeling the strangeness of a rainbow for
the first time, understanding how old aboriginal tales of a rainbow
serpent or deifications of this shimmering vastness might come to
be. On a boat ride out of Ólafsvík in search of orcas, I remember
looking up and seeing, for the first time in my life, the curvature of
the earth's atmosphere, the undeniable dome above us all; later, I
would think again of Flammarion's fantastical engraving.

The weather changed quickly from hour to hour as we drove
across the landscape of moss-covered volcanic rocks and yellow-
ing clumps of grass dotted with horses with modelesque manes
and muddy streaks of ice; the clouds moved with remarkable fluid-
ity, and they sometimes hovered right above the ground in a kind
of primordial mist. As someone who had grown up on a tropical
island in the Caribbean, I was accustomed to the way that weather
can change suddenly from one side of an island to another, but this
felt amplified in Iceland, as if to remind me that the weather, like
the sky, was a living thing, ever-in-shift, as Heraclitus's river. I felt,
distinctly, that I was experiencing the deep presence and majesty
of the sky as never before. I felt like we were drifting through a
landscape and skyscape of pure mythopoeia, a world whose air
was filled with old stories.

I felt this sense of wonderment the most, though, on a blister-
ingly cold night in Iceland, when I stood on the side of a street in
Þkkvibær surrounded by farmland and stared up at the night in
awe, for the aurora borealis had decided to dance across the dark
in green and gray-white. For almost an hour, my wife, I, and a star-
tled group of East Asian tourists who had just arrived at the same
Airbnb we had been at for days gazed at the sublime, shimmering

celestial show despite the boreal winds that ripped at us. Each time we took off a glove to try to take a picture on our phones—my gloves' touchscreen functionality was too mercurial to trust—our fingers throbbed from the cold.

The aurorae felt alive. As I reflected on them the next, much cloudier day, I realized again that the entire Icelandic sky also felt animated, in part because the landscape—with its paucity of trees and, away from Reykjavik and the towns, a dearth of any tall buildings—was often so wide and flat that you could see the sky in all its immense glory, its protean weather and fast-moving, low-to-the-ground clouds and spectra of color easy to discern as your eyes swept along the giant dome in the air. And yet, no matter how much of the remarkable skyscape we saw, I always had the sense that there was more to look for, mysteries hidden in the air and beyond, as in Flammarion's engraving.

Over time, I realized that there was no reason to think I needed to be in this Nordic region to appreciate these celestial qualities. The sky was *always* a marvelous, morphing thing, in Iceland or partly hidden behind New York's skyscrapers. It wasn't marvelous only during resplendent sunsets or aurora borealis dances; it just *was*. But few people in my day-to-day life—especially in New York—ever glanced up unless there was rain or thunder or an unusually intense sunset, and that would always be a momentary glance at best. The ground was where people lived and looked. (Not that most of them paid much attention down there, either, eyes focused more on social media apps on their phones than on the details of the vast, marvelous landscapes all around them.) They lived, I found myself thinking, dividing sky from ground, one worth looking toward only in exceptional moments.

This tendency to ignore the sky reflects a wider truth about air, if we think of it as one of the four classical elements. Because we can't easily see it, we tend to think of it less than the other, more obvious elements, and because so much of our life takes place on the ground, there is a kind of simple evolutionary sense in focusing on the elements that most directly affect us. Of course, the air and the heavens affect us, but they feel more distant, at least to me, than the immediacy of fire or the blue depths of water or the earth beneath my feet. Because so much takes place on the earth, it becomes instinctual to live with your mind there, on something more overtly tangible, rather than pausing to look up or focusing on the feel of the air. To focus on air, even the vastness of the ubiquitous sky, can feel like an additional step, a deliberate pause, a dreamer's luxury of sorts.

I didn't want to live like that anymore, though, face glued to a phone or to the ground more readily than to the immensities just above. I wanted, I realized, to live with more awe of the mundane and marvelous alike, such that the line between the two begins to vanish. Like Flammarion himself, I wanted to remember just how marvelous the world of the air could be, even if from the ground rather than one of his balloons.

And like his legendary monks, but without the religious symbolism, I wanted to bring the sky down to the ground. I wanted to never forget how remarkable the secular heavens are, how they, too, are always in beautiful flux, even if it may require a more concerted look to see this in the drifting of clouds, the vagaries of wind, the multifarious hues of the atmosphere. The aurorae, as they shimmered, had reminded me to look up and revel—and, even though I am an ocean away back in New York, I still look up, in awe of the air even from the ground.

Revontulet, or foxfire, the ancient Finnish called the green-white flames that dance in the sky on dark clear nights, named, as the legend goes, for the idea that the phenomenon of the aurora borealis—that vulpine viridescence—was formed long ago by a great fox whose tail, lit by a fey glow, let little sparks into the sky as it brushed by branches or bushes or snow. And there is, indeed, a kind of poetic truth in this myth; the largest patch of aurorae we saw looked, indeed, like a great streaking tail spread across the sky, its edges shifting and spiking. It is difficult not to compare their terpsichorean motions to fire, as if they were greenish flames traveling across an invisible log.

But there is an airiness to them, as well, that makes the aurorae difficult to classify in a rigid demarcation of the elements. Created by charged particles rushing into the magnetosphere on solar wind, the colors of the aurorae reflect various elements interacting with the atmosphere. The aurorae also have a curious quality of simply being an extraordinary, ineffable apparition in the heavens, living and yet not alive in any conventional sense, unfurling and expanding at random. You can call them swaying flames, but they also give off the impression of being blown by some invisible wind, like streaks of some ethereal, gaseous thing rippled by a zephyr. They are fluid both in motion and definition, blurring the boundaries of simple categories. There is a sense that you are seeing a kind of cosmic mystery, that you, like Flammarion's crawling explorer, have peeked into a special realm for a bit.

This difficulty of defining what aurorae are like might frustrate someone who requires strict labels, but I love their elemental uncertainty, their essential amorphousness. If there is an inherent quality to the cosmos, for me it would be flux, as if we are all drifting along a river that encompasses the universe. All things are part of this ceaseless flow, the divisions between one thing and other fading by virtue of everything, regardless of origin or identity or shape or form, being part of this river-of-all. Although labels are useful for differentiating one thing from another, I'm no great fan of division or rigid binaries. I prefer thinking in terms of spectra of experience and identity, prefer viewing things as gradual, messy transitions where one thing fades into another, rather than defining things simply as one definite state of being or another. The aurorae seem to exemplify that impulse, their labile strangeness difficult to grasp in words, their appearance always in quiet flux. They may not live in the same way we would define an arctic fox as alive, yet their phantasmal movements *feel* vital, feel like a re-minder of the way that the sky, as a whole, is engaged in its own ever-shifting dance, if more slowly than the legendary lights.

The aurorae, then, show us how to bring the sky down, closer to our heads, if not as literally as in Flammarion's fable. They remind us of the heavens' cosmic bizarrerie, and in turn, our own. They remind us of the ethereal, ineffable division between the living and the nonliving, of the fact that life and nonlife are all interwoven if we stop instinctively trying to divide them as categories.

And indeed, when I saw them on that blustery gelid night, I felt a kind of kinship with those luminous cloudflames. I imagined my body slowly rippling at the edges, ever in motion. *What a thing*, I wrote of them in a note with the unapologetic flamboyance of Flammarion, *to become aurorae, shimmering and shifting at the edges, brief and mar-velous and protean! How like humans they are in this description! What a mirror to look into, yet so few realize the auroras are a mirror at all!*

They are like us, you see, those *revontulet*, a way to visualize an aspect of our life in the grand rushing mystery. They reflect our own brief dance on an inscrutable stage of star and void. They shift, always changing, like a quick reel of a human life replayed abstractly on the screen of the sky; then they fade, brief vulpine candles out, and there is just the dark and the stars, the Lights gone until, perhaps, they fox-trot again somewhere else for a little.

But what a thing to dance for a bit, before I return to the fath-omless blank of the night.

Nearly two months later, my wife and I were sitting on a bench near our apartment in Queens as the sun slipped below the horizon. We had each taken a dose of mescaline that morning, which took the form of tall glasses of strikingly bitter green sludge; for hours on the gentle drift of the cactus' dreams, all things had become a bit brighter and more marvelous, something I have become accus-tomed to—to the extent you ever can be—with psychedelics. After a long walk during which we pointed out telephone poles, crenellated brick walls, and the patterns of shrubbery as if we had never seen them before, we had returned home to lie down and listen to music. But, as the sun began to set, I proposed slipping back outside one more time to sit together on that wooden bench.

The dusk air was crisp against our skin, and though New York can often be a discordant symphony of sirens and shouts, all I heard was the soft susurrus of the wind, along with the occasional *whoosh* of an electric scooter zipping down the road by our apartment building. By now, the intense brightness with which the San Pedro had seemed to light all things had faded into a calm afterglow, as had the world around us.

We sat until the sky had put on her indigo, a few stars glittering in her hair. It was at once quotidian and sublime. The buildings blocked some of our view, as is unavoidable in New York, but it was lovely, all the same.

It didn't matter what air surrounded us, I realized, be it Icelandic or that of the Empire State. The wonder I'd felt in that Nordic country was just as valid here; the air, the sky, was just as miraculous if you took the time to feel it against your skin, to look up at the heavens. I no longer needed the aurorae or special vastness of vistas to feel that the sky here, too, seemed like a living thing, full of motion and mystery and myths, whether ancient or yet to be told.

How wonderful, I wrote after we had returned inside, once again channeling Flammarion's style, to find a bit of heaven not just in a distant world of fungus-flecked rock and ice-lined cascades, or when those foxfires dance, or in a mad monkish quest, but in any bit of sky you might find yourself under.

notes

1. Richard Homes, *Falling Upwards: How We Took to the Air: An Unconventional History of Ballooning* (New York: Vintage, 2013), 228.
2. Homes, *Falling Upwards*, 231–32.

Permissions

These credits are listed in the order in which the relevant contributions appear in the book.

"Bodies in the Air" by Aimee Nezhukumatathil originally appeared in *The Kenyon Review*.

"Saturn's Rings" by Ellen Bass, from *Like a Beggar*. Copyright © 2014 by Ellen Bass. Reprinted with the permission of The Permissions Company, LLC on behalf of Copper Canyon Press, coppercanyonpress.org.

"A Small Needful Fact" by Ross Gay originally appeared in Split This Rock's *The Quarry: A Social Justice Poetry Database*, https://www.splitthisrock.org/poetry-database/poem/a-small-needful-fact.

"Ozone" by Rita Dove, from *Grace Notes* (New York: W.W. Norton, 1989). Used with the author's permission.

"American Atmosphere" by Craig Santos Perez previously was published in *Venti Journal* 1, no. 3 (Fall 2020).

Acknowledgments

O ur gratitude runs deep for the community of kin who made this series possible. Strachan Donnelley, the founder of the Center for Humans and Nature, was animated and inspired by big questions. He liked to ask them, he enjoyed following the intellectual and actual trails where they might lead, and he knew that was best done in the company of others. Because of this, and because Strachan never tired of discussing the ancient Greek philosopher Heraclitus, who was partial to Fire, we think he would be pleased by the collective journey represented in *Elementals*. One of Strachan's favorite terms was "nature alive," an expression he borrowed from the philosopher Alfred North Whitehead. The words suggest activity, vivaciousness, generous abundance—a world alive with elemental energy: Earth, Air, Water, Fire. We are a part of that energy, are here on this planet because of it, and the offering of words given by our creative, empathic, and insightful contributors is one way that we collectively seek to honor *nature alive*.

A well-crafted, artfully designed book can contribute to the vitality of life. For the mind-bending beauty of the cover design, cheers to Mere Montgomery of LimeRed; she is a delight to work with and LimeRed an incredible partner in bringing to visual life the Center for Humans and Nature's values. For an eye of which an eagle would be envious, a thousand blessings to the deft manuscript editor Katherine Faydash. For the overall style and subtle touches to be experienced in the page layout and design, we profoundly thank Riley Brady. We also wish to thank Ronald Mocerino at the Graphic Arts Studio Inc. for his good-natured spirit and attention to our printing needs, and Chelsea Green Publishing for being excellent collaborators in distribution and promotion.

Thank you to our colleagues at the Center for Humans and Nature, who are elemental forces in their own rights, including our president Brooke Parry Hecht, as well as Lorna Bates, Anja Claus, Katherine Kassouf Cummings, Curt Meine, Abena Motaboli, Kim Lero, Sandi Quinn, and Erin Williams. Finally, this work could not move forward without the visionary care and support of the Center for Humans and Nature board, a group that carries on Strachan's legacy in seeking to understand more deeply our relationships with *nature alive*: Gerald Adelmann, Julia Antonatos, Jake Berlin, Ceara Donnelley, Tagen Donnelley, Kim Elliman, Charles Lane, Thomas Lovejoy, Ed Miller, George Ranney, Bryan Rowley, Lois Vitt Sale, Brooke Williams, and Orrin Williams.

—**Gavin Van Horn and Bruce Jennings**
series coeditors

Daegan Miller sends his deepest thanks to the writers, poets, and editors who breathed life into *Air*; to all the other elements, without which *Air* means nothing; to the Jalopies for their unfailing commitment to inspiration; and to the Wild Ones of Ashfield, whose laughter makes the world good.

Contributors · volume ii

Báyò Akómoláfé (he/him), PhD, rooted with the Yoruba people in a more-than-human world, is the father to Alethea Aanya and Kyah Jayden Abayomi, the grateful life partner to Ije, son, and brother. A widely celebrated international speaker, posthumanist thinker, poet, teacher, public intellectual, essayist, and author of two books, *These Wilds beyond Our Fences: Letters to My Daughter on Humanity's Search for Home* (North Atlantic Books) and *We Will Tell Our Own Story: The Lions of Africa Speak*, he is the founder of the Emergence Network and host of the postactivist course-festival-event We Will Dance with Mountains.

Darran Anderson is the author of *Imaginary Cities* (University of Chicago Press) and *Inventory* (Farrar, Straus & Giroux). He is an Irish writer, currently living in London.

Sohini Basak (she/her) is a writer and independent books editor from Barrackpore, India. Her first poetry collection *We Live in the Newness of Small Differences* was awarded the inaugural International Beverly Prize and published in 2018. She studied literature and creative writing at the universities of Delhi, Warwick, and East Anglia, where she received the 2015 Malcolm Bradbury Grant for Poetry. She has recently contributed to *Speculative Nature Writing: Feeling for the Future* and is working on her first novel, *The Hospital for Plants.*

Ellen Bass's (she/her) most recent poetry book is *Indigo* (Copper Canyon Press, 2020). In 1973 she coedited the first major anthology of women's poetry, *No More Masks!*, and in 1988, she cowrote the groundbreaking *The Courage to Heal.* Among her awards are fellowships from the Guggenheim Foundation and the National Endowment for the Arts, and four Pushcart Prizes. A Chancellor Emerita of the Academy of American Poets, Bass founded poetry workshops at Salinas Valley State Prison and Santa Cruz, CA, jails, and she teaches in the MFA writing program at Pacific University.

Sara Beck (she/her) is a mother, yoga teacher, life coach, and writer living in her hometown of Fort Wayne, Indiana. Her weekly newsletter *Notes from the Heartland* investigates the intersection of body and spirit as it plays out in the challenges, joys, comedies, beauty, and questions of everyday life. Other work can be found in publications such as the *New York Times, Nowhere, Literary Mama,* and the *Normal School.*

Gabrielle Bellot (she/her) is Staff Writer at Literary Hub. Her work has appeared in the *New York Times, New Yorker, Paris Review Daily, New York Review of Books, Guernica, The Atlantic,* and many other places. She lives in Queens, NY, with her wife and is at work on an essay collection.

Rita Dove (she/her) won the 1987 Pulitzer Prize for her third book of poetry, *Thomas and Beulah*, and served as US Poet Laureate from 1993 to 1995. She is the only poet to receive both the National Humanities Medal and the National Medal of Arts. Recent honors include the Wallace Stevens Award, the American Academy of Arts & Letters' Gold Medal in poetry, the Ruth Lilly Poetry Prize, and the Bobbitt Prize for lifetime achievement from the Library of Congress. Dove teaches creative writing at the University of Virginia; her latest poetry collection, *Playlist for the Apocalypse*, appeared in 2021.

Ross Gay (he/him) is the author of four books of poetry: *Against Which*; *Bringing the Shovel Down*; *Be Holding*, winner of the PEN American Literary Jean Stein Award; and *Catalog of Unabashed Gratitude*, winner of the 2015 National Book Critics Circle Award and the 2016 Kingsley Tufts Poetry Award. His first collection of essays, *The Book of Delights*, was released in 2019 and was a *New York Times* bestseller. His new collection of essays, *Inciting Joy*, was released by Algonquin in October 2022.

Benjamin Kunkel (he/him) is the author of *Indecision*, a novel; *Buzz*, a play; and *Utopia or Bust*, a collection of essays. He was a founding editor of *n+1* and sits on the editorial committee of *New Left Review*. He lives in Colorado.

Antonia Malchik (she/her) has written essays and articles for *Aeon, The Atlantic, Orion, High Country News,* and a variety of other publications. Her first book, *A Walking Life,* is about walking's role in our shared humanity and the damages caused by a car-centric culture. Antonia lives in northwest Montana, where she is working on a book about ownership and writes *On the Commons,* a newsletter about the commons, private property, commodification, and humanity's relationship with ownership, ecosystems, and one another.

Daegan Miller is an essayist and critic, and the author of *This Radical Land: A Natural History of American Dissent.* He lives with his family in the hill towns of Western Massachusetts. You can find out more about Daegan at daeganmiller.com.

Aimee Nezhukumatathil (she/her) is the author of the nature essay collection *World of Wonders* and four books of poems. She is Professor of English in the MFA program at the University of Mississippi and poetry editor of the Sierra Club's *SIERRA* magazine.

Craig Santos Perez is an indigenous Chamoru from the Pacific Island of Guam. He is the author of six books of poetry and the coeditor of seven anthologies. He is Professor in the English department at the University of Hawai'i, Manoa.

Roy Scranton is the author of *Learning to Die in the Anthropocene*, *War Porn*, and other books. He teaches at the University of Notre Dame, where is the founding director of the Environmental Humanities Initiative.

Nicholas Triolo (he/him) is a writer and long-distance mountain runner from Missoula, Montana. With an MS in environmental writing from the University of Montana (2016), Triolo was formerly Digital Strategist for *Orion* and is currently Senior Editor for *Outside Run* and *Trail Runner*. His writing, films, and photography have been featured in *Orion*, *Dark Mountain Project*, *Outside Online*, *Best American Poetry Blog*, *Terrain.org*, *Patagonia's Dirtbag Diaries*, *Juxtaprose*, the *Wild and Scenic Film Festival*, and others. He's currently working on a book, *The Way Around*, about walking in circles, due in fall 2025 from Milkweed Editions.

Michele Wick (she/her), a writer and Lecturer in the Psychology Department at Smith College, studies the human dimensions of the climate crisis, particularly the impact of art on climate related beliefs, behaviors, and human flourishing. She is cofounder and cochair of *Arts Afield*, a program that fosters dialogue across the arts, humanities, and sciences at Smith's Ada and Archibald MacLeish Field Station. Her writing, often a mixture of personal experience and research, can be found at the Center for Humans and Nature and her blog *Anthropocene Mind* (*Psychology Today*).

Andrew S. Yang (he/they) works across the visual arts, natural sciences, and expanded research to explore our earthly entanglements. His projects have been exhibited from Oklahoma to Yokohama, including the 14th Istanbul Biennial, the Museum of Contemporary Art Chicago, and the Smithsonian Museum of Natural History, with curatorial projects *Earthly Observatory* and *Making Kin—Worlds Becoming* for the Center for Humans and Nature. His essays appear in *Leonardo, Art Journal,* and *Kinship: Belonging in a World of Relations* and are forthcoming in the *Routledge Handbook of Visual Arts Practice.* Yang is Lash Chair in Environmental Education & Sustainability at Hampshire College.

Felicia Zamora (she/her) is the author of six poetry collections, including *I Always Carry My Bones,* winner of the Iowa Poetry Prize and the 2022 Ohioana Book Award in Poetry. She's received fellowships and residencies from CantoMundo, Ragdale Foundation, and Tin House. She won the 2022 Loraine Williams Poetry Prize from the *Georgia Review* and the 2020 C. P. Cavafy Prize from *Poetry International.* Her poems appear in *Best American Poetry 2022, Boston Review, Guernica, Orion, The Nation, Poetry Magazine,* and others. She is Assistant Professor of Poetry at the University of Cincinnati and Associate Poetry Editor for *Colorado Review.*